```
AAA      CCC  U   U  BBBB  III  K   K    Y   Y  EEEE  AAA   RRRR
A   A    C   C U   U  B   B  I   K  K      Y Y   E      A   A  R   R
A   A    C      U   U  B   B  I   K K        Y Y   E      A   A  R   R
A   A    C      U   U  B   B  I   KK         Y Y   E      A   A  R   R
AAAA    C      U   U  BBBB  I   K           Y    EEEE  AAAA  RRRR
A   A    C      U   U  B   B  I   KK         Y    E      A   A  RR
A   A    C      U   U  B   B  I   K K        Y    E      A   A  R R
A   A    C   C U   U  B   B  I   K  K        Y    E      A   A  R  R
A   A    CCC  UUU   BBBB  III  K   K        Y    EEEE  A   A  R   R
```

BY

HONABE

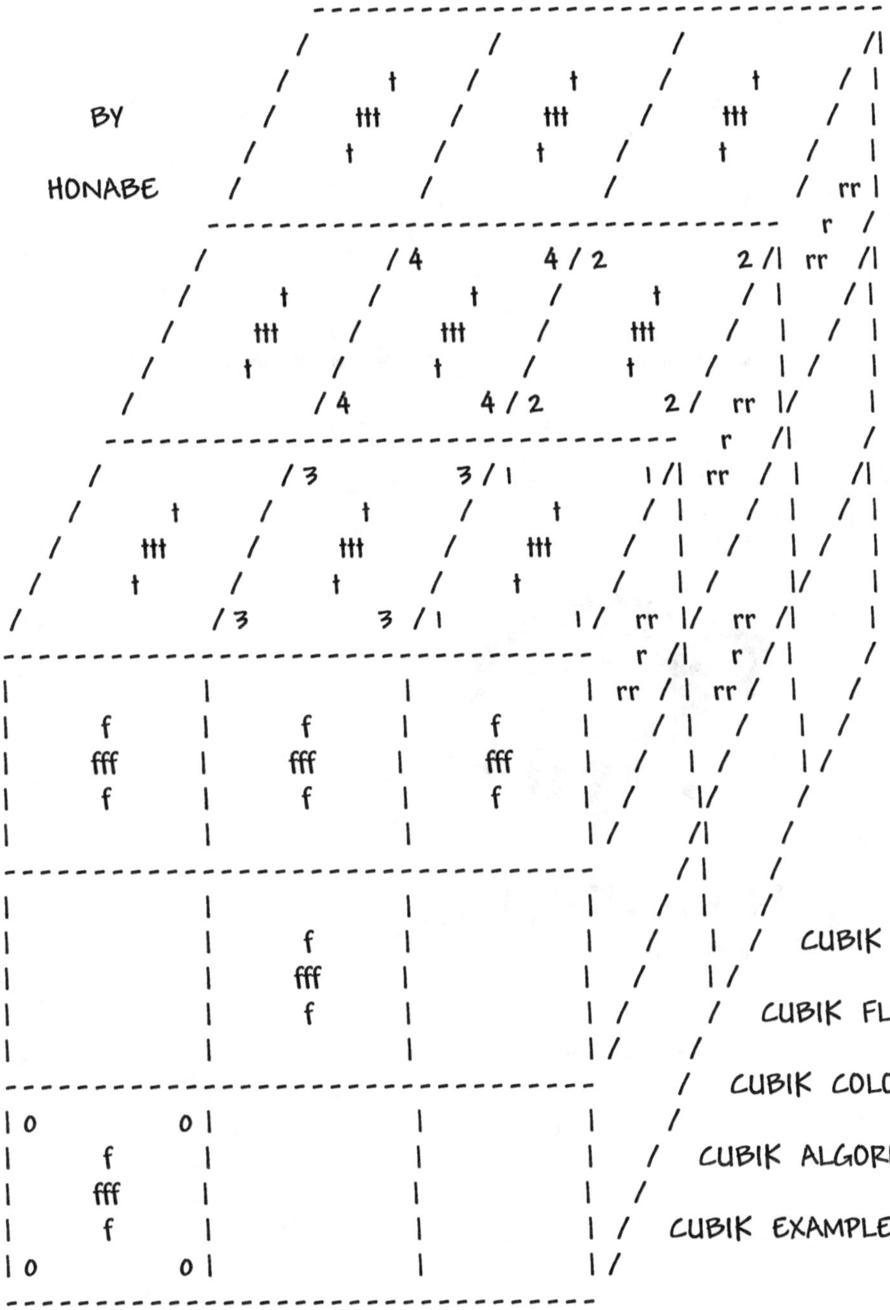

CUBIK NOTE BOOK

CUBIK FLOWCHARTS

CUBIK COLORING BOOK

CUBIK ALGORITHM TABLE

CUBIK EXAMPLES (let's do it!)

@ 2002 HONABE 'A CUBIK Year'

First published by AuthorHouse 06/04/04

ISBN: 1-4184-5479-6 (e)
ISBN: 1-4184-5480-X (sc)

Printed in the United States of America
Bloomington, Indiana

This book is printed on acid-free paper.

This book is the result of an unpublished 1982 manuscript converted to Microsoft Word 6 and is updated appropriately. It's been 1982 all over again!

authorHOUSE
Your Voice in Print

1663 LIBERTY DRIVE
BLOOMINGTON, INDIANA 47403
(800) 839-8640
www.authorhouse.com

CONTENTS

Preface

* * * Introducing Twenty Years of Cubistry * * *

Twenty years of cubistry. Wow! That is twenty years that have gone by and in the last ten years I have enjoyed giving a Rubik's CUBE to each of my daughter's families so the grand children are able to be acquainted with them (also gave them each a 2 X 2 pocket CUBE). Whenever I see any of them scrambled I always enjoy unscrambling them but intuitively it is very probable that they prefer them to be left scrambled. A few times while at my daughter's in laws or at some friends homes and I see a scrambled CUBE lying about, I'll pick it up and retreat to a corner and unscramble it. My daughter then would comment, "Dad, you haven't lost your touch". Often these CUBES are presents that the wife buys their husband for their entertainment (what a presumptuously form of entertainment). It is our family doctor that recently keeps my cubistry interest since once a month I take my wife there for a shot and he has a CUBE sitting on his desk along with a lot of other interesting paraphernalia. I always pick it up and take it to a chair near his waiting area and I proceed to make him a new pattern and where appropriate reveal to him some of its mechanical limitations. This has been happening for some three years now. Besides, when his patients often see me tinkering with the CUBE off to the side they all in some way jester and say with a chagrin that "I played with that thing". It is such a puzzle that that is all anyone seems to say about it. I also always keep a 2 X 2 CUBE with its little net hanging from a lamp in our family room. Often it is tinkered with but so far no one has come close to unscrambling it.

A recent engineering magazine had the results of a robotic contest and it included solving the Rubik's CUBE. It used a digital camera, a computer, and a servo type mechanism for holding and twisting the CUBE, thereby solving a scrambled CUBE in twenty to thirty moves. However, the servo type mechanisms are so slow that I'm certain me and my 'cubik' solution could beat it, time wise! Soon after reading this article I happened to peer into the inside cover of my 1982 Taylor/Rylands pattern book and there was pasted the following news clip, quote:

"Puzzle solver:
Ronnie Rubik, a robot that can reportedly solve the Rubik's Cube puzzle in less than 0.2s, has been built by engineering students at the University of Illinois at Urbana-Champaign. Although the robot "knows" how to achieve the solution almost instantaneously, its solenoid-driven hands can manipulate the cube at a speed of only 12-moves a minute. Built at a cost of $300, the robot includes a color-sensing visual system, a computer brain, and mechanical hands.
-Robert B. Aronson"

Two other events happened. One is when I am downtown I sometimes stop in and survey puzzles in a game-hobby shop. About five years ago a 5 X 5 CUBE got my attention so the next time I saw it I purchased one. However, it is still in its plastic case in its pristine state. A couple of times I told Bill about it (he is in my 'cubik' story) but I don't think he ever purchased one of them. The other was for some unknown reason I bought myself a "Get-back-to-square-1" puzzle for this past Christmas. Well 'dummy' me I took it out of the box without heeding to its suggestion to see inside before twisting and turning it. The more I twisted and turned it the faster I realized that this was quite some puzzle. Then it dawned on me that it seemed familiar. When I retired in 1992 'in jest' they gave me a box of puzzles at my retirement luncheon. When I went upstairs to look in the box, there it was, another 'Square-1' puzzle. The more I tried to use 'my expertise' to solve it on my own, the more hopeless it became. I then realized the only hope I have is to make it look like the configuration in the unopened box especially after reviewing the instructions from the first box. That puzzle sitting in the unopened box was definitely an immeasurable help. I decided to arrange the eight small triangles first, then do the eight four sided pieces. Well, it was somewhat easy to arrange four of the triangles on one plane, but the other four on a different plane was not easy to come by. The better part of a day I tried to deduce some strategy but it was to no avail, however, I did improve upon the time it took to mess it up and start from 'scratch' again. Often two of the last four triangles would pop up okay, but that was all. So the next evening I reasoned if I kept messing it up I should eventually go through a sufficient number of iterations until by chance all eight triangles popped up okay and that is what eventually happened, PHEW! I then used the same iterative strategy on the eight four sided pieces being very careful not to disturb the eight triangles. Eventually the eight four sided pieces popped up okay and at least that didn't seem to take as long a time as the eight triangles. Then following the instructions, I got back to square one and I have no desire to do it again.

Back in 1982 or the early eighties a lot of Rubik's puzzles became available. I remember displaying some of them on our family room bookshelf. There were my three 3 X 3 CUBES with different patterns, two 4 X 4 CUBES (one with a pattern), a 'TOMY'- MEGAMINX-Dodecahedron, a 2 X 2 pocket CUBE, and a Pharaoh's Dilemma. Over time my teenage daughter's friends (both boys and girls) would tinker with them (make patterns) and in most instances I could recover them, however, I was particularly concerned about the 4 X 4 and the MEGAMINX. Eventually one of the 4 X 4's and the MEGAMINX became unrecoverable. I put the MEGAMINX aside and concentrated on recovering the 4 X 4. Things about the 4 X 4 appeared similar to the 3 X 3 and I reasoned to try to align the two inner slices of my RUBIK's REVENGE. In order to do this I generated a few good notes. Of course, once the inner slices are in alignment, the 4 X 4 then solves as if it were a 3 X 3 CUBE. After recovering the 4 X 4 three or four times I passed the word to my daughters that they were quite difficult to recover (took a lot of time). After that they weren't disturb as much. Doing a MEGAMINX recovery entailed many copious notes over several days. Once it was unscrambled I hid it away, I didn't want to spend the time it took to do it again. Of course the 2 X 2 solves like solving just the corner cubes of the 3 X 3.

The other Rubik's puzzles that I purchased as collectors items for they are still in their unopened boxes are: Rubik's World, Alexander's Star, Rubik's Clock, and Rubik's Magic (Link The Rings & Create The Cube). My grandson and I tinkered a little bit with Link The Rings. I didn't happen to purchase any of the cubic type puzzles that apparently were the Rubik's CUBE mechanism in disguise being in an asunder of shapes. Then there is a key chain. A Rubik's CUBE on a key chain! I still have one of them hanging on a hook in my train room. It was neat to have one of them in our car when I was waiting for my wife or daughters for when I had gotten a Rubik's brainstorm I could try it out which I often did wearing those stickers off a couple of the cubes.

That's the thing about the Rubik's CUBE, you either wore the stickers off the 'cublets' or after a lot of wear and tare it started coming apart frequently when catching an edge while twisting it. Of course after a certain amount of wear it was much easier to take it apart and then reassemble it again. I wore out at least a couple of 3 X 3's, one 4 X 4, and two 3 X 3 key chains, that's a lot of twisting. I still have all the pieces in a bag for curiosity purposes. Also I read somewhere that some 'cubist' lubricates them in order to make them easier to twist which must also reduce the wear and tare.

When my manuscript was nearing completion I remember putting a middle size version of the Rubik's Revenge on the tube at work during my lunch hour. I gave some thought of having a Revenge addendum to my Cubik's story but I really didn't as yet have a real handle on some 4 X 4 algorithms to align the two inner slices. Besides, I still wasn't that happy overall with my set of 3 X 3 algorithms. When I come to think of it, over the last twenty years it seems I have been all along mentally updating my algorithms and flowchart. But it is such a interesting challenge to just to use word processing to display my Cubik Flowchart and Cubik Solution. Of course another thing is that over the last twenty year, other than Bill, I have not ran into one other single person much interested in solving the CUBE. It must be an odd thing to want to become interested in but recently my son in law, with me watching over his shoulder, was surfing the net for a publisher, we also took a brief cursory look at the Rubik's CUBE sites so there is an interest there.

Soon after every Thanksgiving we have our usual reunion (pizza party) at one of the retiree's home. Usually ten to fifteen retirees attend along with ten to fifteen engineer/technicians still working in our engineering section. It would certainly be neat to hand out, finally, a copy of my book, especially if it is a total surprise.

* * * Forward To Publishing * * *

When I retired in the spring of '92 due to downsizing, my thoughts were on how to keep busy and how to make it to early Social Security. I told an engineering friend (for thirty-five years, a skiing buddy), "Dave-the-1st", that I have made up my mind that on Monday mornings I'm going to make believe I'm going to work. Park my car in the company lot, walk three squares and have eggs and grits, walk another three squares to the Ben Franklin Bridge, walk across it (about two miles long) to Philadelphia, tour the sites for a couple of hours, and return on the ferry. (Philadelphia is a beautiful site with the morning sun at our backs illuminating the city like a city of brilliant jewels.) It seemed to be a neat idea. My friend seemed interested and later decided to join me. To this day his wife never understood why we were doing this or even why people want to go skiing for that matter. My wife still had a twenty-seven hour part time job. At that time our state had an eight to nine percent unemployment rate so we were eligible for assistance so once a week we all used a company sponsored site to honor our commitments, take some interviews, etc. (it was like a mini-reunion). Besides interviewing I had my sights on perhaps a part time courier job but that fall except a part time job (one day a week to see if I liked it) with a local newspaper handling complaints in South Jersey. Well, it was a real fun job. For four hours seeing towns and places I have never been through before even though I have spent my whole life in the area. Soon the manager asked if I could use some more time and I complied but I still wanted my Mondays for the bridge. He really wanted five mornings a week but lucky for me he was able to get one of his weekend workers to cover for me on Mondays. For two solid years "Dave-the-1st" and I walked the bridge (if snow or ice closed the walkway we took the ferry both ways) until he bought a thirty-five acre farm in New Hampshire in order to be near his sister. So, no more walking the bridge and I went to five mornings a week for the next three years until I retired-retired to help my daughter and son in law to fix up a house they just bought and my wife started to wind down as she approached sixty-two for we wanted to travel. For the next four years we traveled aplenty and I walked the bridge no more than five or six times each year, alone. It's annoying, it seemed everybody knew I walked the bridge (a one mile up hill climb by the way), several expressed interest to join me (even my wife and her friends) but no one ever did. Once a neighbor was to come with me but called 6:30 AM that morning and cancelled.

Last fall the bridge was closed to pedestrians due to 9/11 security reasons. A week or so before Christmas "David-the 2nd" and I happened to notice the same article in our local newspaper indicating they were going to reopen the walkways. I see him every Thursday morning at a 7AM juice-coffee-pastry-study hour of which he sometimes leads. Well, that next Thursday he sincerely expressed interest in walking the bridge with me and indeed I was aghast. ("David-the 2nd" is a recently retired teacher who is a history, architecture, and artist buff and has hosted several recent trips to Europe. He also enjoys "O" gauge model railroading like myself.) After several weeks of engaging in conversations while walking the bridge regarding the study subject, "O" gauge railroading, and memories (I've known him for some thirty-three years), I discovered that he just became a center city tour guide and was currently involved in starting to publish a series of books. How interesting! "Dave-the 1st" and I must have walked up and down in center city every main street, every side street, every alley way, etc. all the time reading all the names on all the buildings plus reading what is on all those plagues. Now for "David-the 2nd" he is acquainted with all those names, etc. Wow, he is connecting everyone of those dots foe me so to speak. Not only does he foremost wants the exercise, but is looking for interesting places to take his tour groups when he needs to fill in time between site appointments/scheduling. He has publishing encouragement since two of his neighbors near his summer home at the Jersey Shore are associated with the literary agency/ publishing business. Over the weeks he gradually aroused my dreams over the fact that under my train platform is an 88-page manuscript and when he alluded to inter net publishing, well that really got my attention.

My two son in laws often use the inter net and perhaps some half a dozen times I would look over their shoulders to gain a little knowledge of this inter netting. When one of them realized I had a manuscript ready to go he did a search and he subsequently E-mailed "Ist Books". After receiving their publishing plan, it took me two weeks to update my manuscript with several annotated improvements. This is a lot of background material but it is meant to serve as a background for perhaps two subsequent books (Lord willing) regarding all the notes I have generated since retiring which pertain to winning streaks in Blackjack (results of playing 125,000 hands using a hand held game; 50,000 more to go) and pertain to my four and half level "027" model train layout recently featuring the automatic sequencing of trains (three of them) and automatic switch control via industrial twelve volt AC relays. (the sequencing is still in a design stage). The latter could be a fifty-five year "027" saga.

      Back in the spring of 1982, my youngest daughter had an Indoor Guard and Wind Ensemble competition in New Port, R.I.. My wife and I met her there and enjoyed the competition and the sight seeing with her classmates plus the delicious seafood. On the way home we spent two nights with the minister and his wife that married us (lived near the Connecticut/Rhode Island border), played with their Rubik's CUBES, and then stopped in to 'check out' a pay-to-publish publisher in New York City since I had my manuscript with me. They perused it and after some time gave me a price, and that was that. After I returned home I more or less mumbled to my dad that I would like to have my manuscript published at a certain fee and he responded by just giving me a Pennsylvanian Dutchmen's glare, that was that, nary a word was said. With my oldest daughter a freshman in college and the other a junior in high school plus after several publishing inquiries, I just stowed it away.

* * * Prelude To Word Processing * * *

      Here I need to acknowledge all those back in 1981/82 who helped me over come my Word Processing hang ups. It was solely my idea to put the CUBE on the TUBE. A fellow engineer in our engineering section created the flow chart graphics. Then to acknowledge my son-in-law for giving me the go ahead to use their computer in order to put a Microsoft-Word-CUBE on the TUBE plus my daughter for holding my hand while over coming lots of Microsoft Word hang ups. At first it seemed to be some kind of a Microsoft Behemoth that was wished on John Q public by Bill Gates and then no wonder the constituency wanted to hang him by his micros so to speak. Well my son-in-law uses Microsoft Word all the time and kept assuring me really how forgiving it is. All this after losing a couple of files by inadvertently hitting CNTRL "A", so he demonstrated emulating my hunt and peck technique. DUMMY ME! Well, it slowly-slowly made me a believer! As I got further and further into my 2002 manuscript, it seemed like it was 1982 again being immersed in word processing.

      My original 1982 small CUBE diagram is two inches high at six lines per inch. The base is 1 ¼ inches wide with the right side at a 58 degree angle determined by the slope of the virgule '/'. For my 2002 CUBE, it was impossible to generate a CUBE using the Times New Roman font since the virgule was way too steep. Then after waiting for my son-in-law to come home from work, he showed me how to surf the zillion fonts (or it seemed to be) but we quickly zeroed in on the font Andy. Then after generating 'ANDY' CUBE, I found it to be 2 ¼ inches high at 5 1/3 lines per inch. The base is a little over 13/8 inches wide with the right side at a 66-degree angle determined by the slope of its virgule. After writing the introduction part using ANDY for my Figures 1 and 2, I became more curious about all the other existing fonts since the right side of "ANDY" had some distortion. My original CUBE had absolutely no distortion, probably because the Diablo 'dot-matrix' print wheel printed all letters (upper and lower case) space bars, and punctuation with the same spacing so even annotating the CUBE had no affect on its absolute symmetry.

Preface

In looking for a better 'cubik' font, I printed out some forty-one of them four of which were possibilities as follows: ROMAN /T/ |t|; ANDY /T/ |t| ; EUROSTILE /T/ |t| ; and NEWS GOTHIC MT /T/ |t| . The slope of the virgule for the NEWS GOTHIC MT font is the same as for ANDY but the EUROSTILE virgule is more gradual than my original Diablo virgule (55 1/2 degrees compared to 58 degrees). After all this, I kept coming back to ANDY basically since the inside width of a cublet is three dashes (---), the same as my original Diablo cublets. NEWS GOTHIC needs five dashes; EUROSTILE needs a 'zillion' dashes. I actually designed a CUBE with NEWS GOTHIC MT and it looked promising but ANDY's three dashes are easier to see and ANDY is a darker font. I started to construct a CUBE using EUROSTILE but soon found the virgule to be too elongated and gave up. When I then compared ANDY in detail with my Diablo CUBE, I then focused on revising my ANDY CUBE to be in agreement with my original CUBE as much as possible and called it ANDY(PLUS). So Figures 1 and 2 are with ANDY and the rest of the manuscript is with ANDY(PLUS). I am not exactly satisfied in making my CUBE with any of these fonts but since there is some distortion let's say that it makes the CUBES look sketchy. Perhaps this will invite readers to get out their CRAYOLA colored pencils or colored "washable" markers (not exactly recommended since the colors can run or bleed but the coloring looks real neat). ANDY(PLUS) has an overall height 12.5% higher and an overall width 15% lower than my original undistorted CUBE. With regards to sketchiness, the unannotated ANDY(PLUS) CUBES are almost perfect (not too sketchy) but become more sketchy when annotating since the ANDY font for 't' and 'f' is narrower than for 'e', 'r', 'b', and 'k'. The other night at band rehearsal, a retired computer engineer musician friend indicated that word processing fonts do exist that equally spaced all the 'type', meanwhile, I'm restricted to Microsoft Word 6.0. He also confessed that he uses Microsoft Word but doesn't like it. "Did you hear that, Bill Gates, you behemoth you!". He also related that separate software does exist just for making flowcharts.

My original manuscript flowchart used an overlay feature for directional arrows. Horizontally it overlaid dashes with greater or lesser signs; vertically it overlaid a 'vertical bar' with a small (v) or a caret. That's another thing, before a fellow engineer construed of a word processing flowchart, his perusal of a Diablo print wheel catalogue found an acceptable font but it included a vertical bar for his flowchart vertical lines (dashes were used for the horizontal lines). My original cover sheet and Figures 1 and 2 used a capitol (I) for a vertical line but the rest of the original manuscript (including the flowchart) used the vertical bar. For this manuscript I initially decided to use lower case (l) for a vertical line. ANDY's (l) is not exactly straight so it contributes to my Cubik CUBE'S sketchiness. *(Then the next week at band rehearsal I found out that a vertical bar exists for Microsoft Word so the very next time on the computer I used the search and replace to replace the (l) with the vertical bar (|). My daughter and son-in-law were surprised but they evidently never had a need for a vertical bar. For that matter, just the other day while walking the bridge with 'David-the-2nd' and during our usual conversations, I discovered that he too didn't know he has a vertical bar with his Microsoft Word even though he has already generated three several hundred page manuscripts. Vertical bar, it's in disguise!)* Of course the caret and small (v) overlay lined up perfectly for my original manuscript and as you can see this is now not the

case. There is now a slight offset but since there is no overlay it doesn't look too bad.
Originally compared to the dash (hyphen), the asterisk was just slightly off center but
was so much off center for the Microsoft Word flowchart that I had to compensate for
the offset by using an equal sign for my flowchart horizontal line as you can see. Then
lo and behold along with the greater than or lesser than sign, I got a free arrowhead;
what a neat enhancement that turned out to be. It takes at least an hour to do a
flowchart sheet, fun!, fun!!, fun!!!

INTRODUCTION (1982)

The past year with the CUBE has been enjoyable. Once our engineering department acquired a word processor, I decided it would be a challenge and it would take some courage to put down on paper a solution to the Rubik's CUBE. Once the CUBE is solved and the solution is not too difficult to repeat, one can say then that the CUBE has to it SIMPLICITY or RYTHM.

The CUBE became enjoyable since after a very short time with it I soon stopped treating it as a puzzle and started to treat it as a fascinating game by doing the same basic movement, translation, operation, or algorithm over and over again until it was solved. I started with the FOUR STEP APPROACH (described in Ideal,s instruction booklet) and sensed the four step rhythm to it while completing one face of the CUBE, namely:

(1)   READY-------------cube that is to be moved to the top face is in a ready position,

(2)   BRING DOWN----part of the top face is brought down to pick up cube that is ready,

(3)   MOVE IN-----------move ready cube laterally for pick up,

(4)   BRING BACK-----top face has added one additional cube towards completion.
(When READY a cube is MOVED IN either clockwise or counter-clockwise so it is correctly oriented).

So with four steps in twisting the CUBE, in most circumstances, a cube can be properly placed in the top face.

I next expanded the FOUR STEP APPROACH to seven steps where the first step brings down one of the four reference cubes shown on the title page, next it is moved about, and the seventh step puts it back.

This seven-step rhythm is:

(1)   BRING DOWN----bring down either reference cube 1, 2, 3, or 4,

(2)   MOVE OUT--------ref. cube is moved clockwise 1Q (quarter) turn laterally,

(3)   BRING BACK-----top layer is restored but minus ref. cube,

(4)   ROTATE------------ref. cube is rotated 1 or 2 Q turns laterally (this is the step that churns the CUBE, if it is eliminated no changes occur),

(5)   BRING DOWN----bring down top as in Step (1),

(6)   MOVE IN-----------move ref. cube laterally for pick up,

(7)   RESTORE TOP----top is restored and cubes in middle layer and/or bottom layer have been translated (churned).

To solve the CUBE four seven step sequences are used which are arbitrarily named R/B/1/1, R/V/1/1, C/V/1/2, and C/B/1/2.

Here in order to interchange two of the bottom four corners cubes, my original manuscript contained a rather complicated nine step sequence that was a deviation on the seven step R/B/1/1sequence and was arbitrarily named R/0 and RR/0. My original solution focused on combinations or slight variations of these four seven-step sequences. Over the years these R/0 and RR/0 operations became more of a slice operation with five basic steps and with six basic steps, respectively, as follows (easier to remember):

<u>R/0 STEPS</u>
(1)   *BRING DOWN----bring down right side,*
(2)   *ROTATE REAR---rotate rear 1Qt to the left,*
(3)   *BRING BACK-----bring back right side,*
(4)   *ROTATE REAR---rotate rear 1Qt to the right,*
(5)   *RESTORE TOP----use FOUR STEP APPROACH.*

<u>RR/0 STEPS</u>
(1)   *BRING DOWN--------bring down right side,*
(2)   *ROTATE BOTTOM—rotate bottom 1Qt to the left,*
(3)   *ROTATE REAR-------rotate rear 1Qt to the left,*
(4)   *ROTATE BOTTOM—rotate bottom 1Qt to the right,*
(5)   *BRING UP-------------restore right top*
(6)   *RESTORE TOP-------use FOUR STEP APPROACH.*

R/O interchanges two adjacent bottom corners where as RR/0 interchanges two bottom corner cubes diagonally.

Solving the CUBE, therefore, entails the following:
(a)   Either a R/0 or a RR/0 operation correctly positions the bottom four corner cubes,
(b)   Up to four R/B/1/1 translations flip around (twist) the bottom corners until they are oriented correctly,
(c)   The R/V/1/1 translation finishes one-by-one the bottom four edge cubes, thereby completing the bottom layer,
(d)   A C/V/1/2 translation positions correctly the four middle edge cubes (in conjunction with the RM/0 operation if necessary),
(e)   A sequence composed of the R/V/1/1 and C/B/1/2 translations are used to finish the CUBE (reorient the four middle edge cubes if necessary).

A single page table illustrates these seven operations/translations (R/0, RR/0, RM/0, R/B/1/1, R/V/1/1, C/V/1/2, C/B/1/2), and one ENDING ALGORITHM (ALGO#1).

Here again in order to position the four middle edge cubes, my original manuscript contained an algorithm consisting of several of the four seven step sequences. Over the years this algorithm operation also became more of a slice operation with three steps as follows and is arbitrarily named RM/0:

<u>RM/0 STEPS</u>
(1)   *BRING DOWN-------bring down right side 2 Qt's (invert it),*
(2)   *ROTATE MIDDLE--rotate middle layer 1Qt to the left,*
(3)   *RESTORE TOP------bring back right side.*

I still use my ENDING ALGORITHM that contains several seven-step sequences.

As a result of these seven translations or operations, a FLOWCHART could be constructed that diagrams a solution for solving the Rubik's CUBE. The flowchart appears at the end of the story.

This article indicates that these seven translations or operations are cyclic, e.g. if you have a solved CUBE (pristine configuration) and repeat the R/0 translation twelve times (108 twists of the CUBE), the solid pattern returns after the 108th twist. Besides, a group of patterns made from a solid CUBE using Taylor/Rylands' algorithms are cyclic in two, another large group are cyclic in three. A pattern section is included at the end of this story. There, some patterns are made from a solid CUBE as slice operations and the four corner cubes are never disturbed, others are built from a scrambled CUBE.

These seven 'algorithms' are highly similar and require no detailed notations since once described they become repetitive, used over and over again. This algorithm approach lags way behind other approaches in that it is difficult to solve the CUBE in less than seven minutes. If it takes about 150 twists for my algorithms and about 30 twists to first do the top layer, this means 2 ½ seconds for an average twist. The good things about this algorithm approach are that the solution is not too difficult to remember (once you learn to align the CUBE properly as described in the article), it is easy to use when doing complicated patterns from a scrambled CUBE, and other than the bottom four corners and the last four cubes (which are solved in pairs), the cubes are solved one at a time (10 of the 20 cubes are done one-by-one), which also helps when doing patterns.

The first time through I really bogged down with details trying to put this solution on paper. Two-thirds the way through I stopped, completed the CUBIK ALGORITHM TABLE and the CUBIK FLOWCHART and then rewrote the article trying to keep the solution as simple as possible. *(My final manuscript was still very sophisticated.)*

I hope you enjoy my 'cubik' story, the flowchart looks great, and don't forget to color in your favorite patterns. Have fun. YEA!, YEA!!, YEA!!!, for Word Processors.

Dear Cubik World (1982),

    It's hard to believe that a year has gone by since I luckily solved the CUBE by playing games with it. The solution focused on doing translations on the four referenced top cubes 1, 2, 3, and 4 (see CUBE in Figure 1) where each translation is somewhat two consecutive four step sequences diagrammed in the instructions that accompanied the purchase of the CUBE made by IDEAL. Consequently, this solution assumes the reader has mastered the four-step approach in positioning and orienting one face of the CUBE as shown in Figure 1. (A review of this four-step approach is included at the end, however.) It's amazing the number of persons who have purchased the CUBE and, in their haste to get their hands on it, have discarded the pretty box along with the tiny eleven-page instruction booklet, oh well.

    I guess the main reason that I decided to try to describe my solution is due to the fact that a fellow employee added a 'word processor' software package to our computer, so please excuse the elongated shape of the cubes since they are generated mainly from three keys, an 'I', an '-', and an '/'. This presentation is in black and white unless your imagination decides to visualize color schemes as it chooses or you tint the cubic figure.

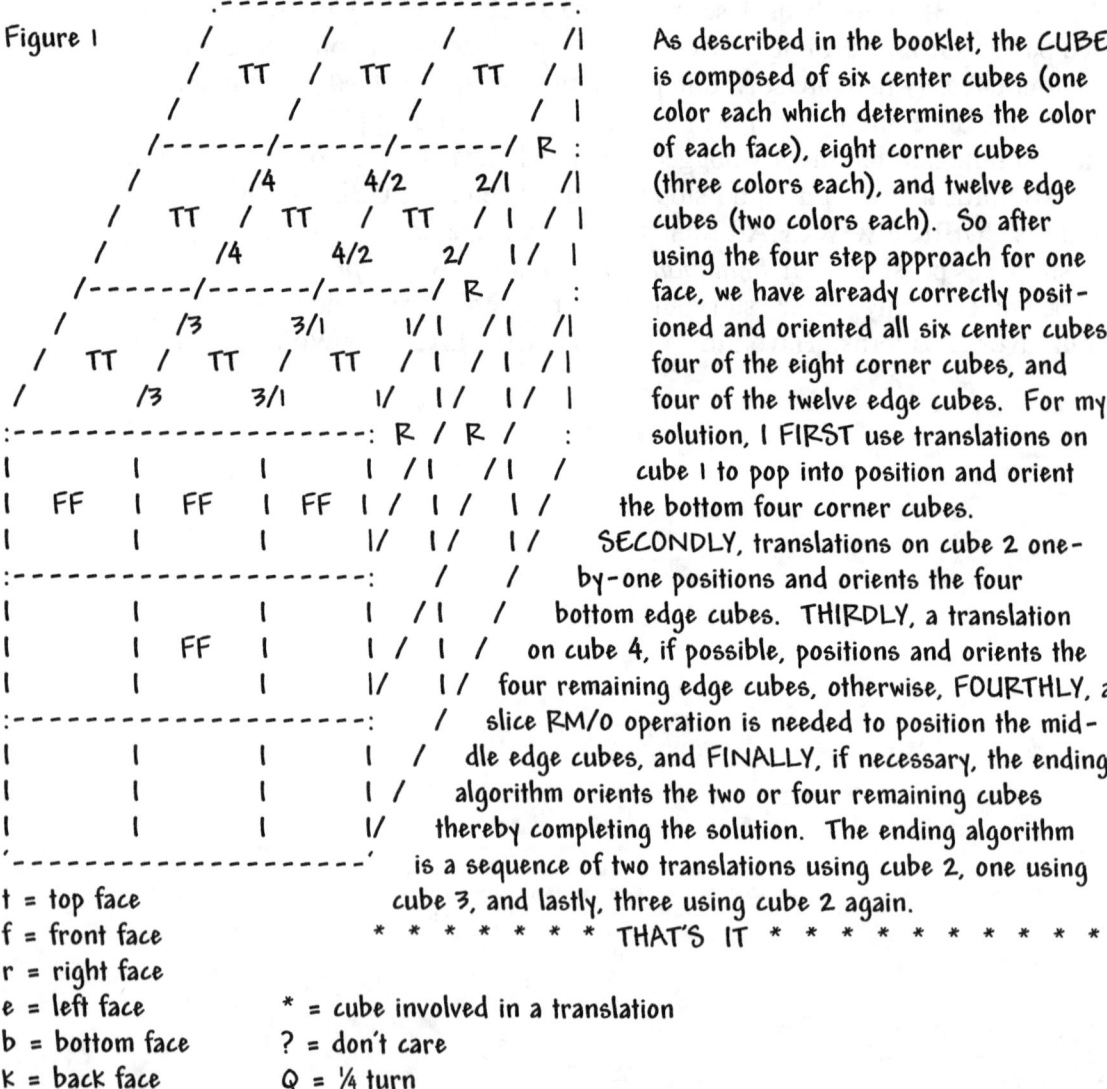

Figure 1

As described in the booklet, the CUBE is composed of six center cubes (one color each which determines the color of each face), eight corner cubes (three colors each), and twelve edge cubes (two colors each). So after using the four step approach for one face, we have already correctly positioned and oriented all six center cubes, four of the eight corner cubes, and four of the twelve edge cubes. For my solution, I FIRST use translations on cube 1 to pop into position and orient the bottom four corner cubes.

SECONDLY, translations on cube 2 one-by-one positions and orients the four bottom edge cubes. THIRDLY, a translation on cube 4, if possible, positions and orients the four remaining edge cubes, otherwise, FOURTHLY, a slice RM/0 operation is needed to position the middle edge cubes, and FINALLY, if necessary, the ending algorithm orients the two or four remaining cubes thereby completing the solution. The ending algorithm is a sequence of two translations using cube 2, one using cube 3, and lastly, three using cube 2 again.

* * * * * * * THAT'S IT * * * * * * * * * *

t = top face
f = front face
r = right face
e = left face
b = bottom face
k = back face

* = cube involved in a translation
? = don't care
Q = ¼ turn

In November and December of '80, I do remember seeing the CUBE being advertised on TV a couple of times and never thought much about it. Near Christmas my wife suggested that a CUBE would be an acceptable present for our younger teenage daughter. I soon found out all the stores were sold out. Also, I remember my dad telling me that he saw them being sold and demonstrated in early December at several of the malls. The toy stores had set up special tables out in front of their stores selling only CUBES for around $15. Anyway, on Christmas day all I could do was to promise my daughter a CUBE as soon as they were restocked. During Christmas week we visited my wife's brother's family and I saw that my niece and nephew each received a CUBE as a present. There were several relatives there and since we were all occupied with looking at presents and involved in many conversations I might have given one of the CUBES a couple of twists, but I still didn't give the CUBE much thought. I guess being busy at work I completely forgot about the CUBE for the next six or seven weeks. Then near the end of February on a Friday night we were back at my brother-in law's house for my nephew's birthday. It was just the two families and they had the two CUBES lying around. I picked up one of them (later my nephew picked up the other) and must have twisted that darling for a couple of hours. During the course of the evening my nine year old nephew out shown me in cubistry but I did manage to get all the cubes on one side the same color and tried for two sides but didn't come close. I inquired if directions came with the CUBE and they said they did but the directions and boxes disappeared during Christmas week probably because they wouldn't have wanted to see them anyway. By now I supposed I was giving the CUBE a little thought.

Early Saturday morning I went to SEARS to pick up my wife's repaired sewing machine and it was on the way home when I remembered that I had promised my daughter that I would buy her a CUBE as soon as the stores restocked them. I immediately headed for a nearby discount toy store and purchased an IDEAL CUBE with the engraved 'Rubik's CUBE' trademark for $4.99 minus tax. Before I left the store I had verified that the box containing the CUBE also contained the little instruction booklet. As soon as I got home I gave the CUBE a twist or two and then started reading the tiny eleven-page instruction booklet. A good feeling seemed to come over me when I read the first sentence, 'Rubik's CUBE is far more than a puzzle....', which made a lot of sense to me because the night before I had no idea what I was doing while twisting and twisting the CUBE trying to get one side of the CUBE the same color. I didn't even surmise the significance of the center cube that has only one face. So after reading about the Four Step Approach, later that day I spent three to four hours confidently positioning and orienting all the cubes on one side as shown in Figure 1. Now as all potential cubist find out, from this point forward you are, without any doubt, on your own. Well, I thought a little about what I was going to do next before retiring for the evening.

The CUBE is a funny entity in that you seem to enjoy playing games with it for around 1 ½ to 2 hours then there is a strong desire to want to lay it aside for a while and go about doing something else. This was the way it was for the next seven days until it just popped out into a solid CUBE.

Late Sunday afternoon I picked it up and came upon my first translation that was one heck of a revelation. When one of the four corner cubes didn't change but the other three did I remembered the fact that the tiny booklet indicated a zillion combinations are possible. Then I reasoned, this could be true yet, in reality, is really not true since a simple translation only moves three, four, five, or six cubes and after repeating the same translation many times eventually you exhaust the combinations until the combination you are looking for appears. This reasoning, however, does not treat the CUBE as a puzzle but as sort of a game using repetitions of simple algorithms (translations). A simple algorithm (translation) based on the Four Step Approach is shown below.

```
Fig. 2A .--------.          Fig. 2B ,--------,          Fig. 2C ,--------,
       /T /T /* /|                 /  /  /K /|                 /  /  /K /|
      /--------/F :               /--------/T :               /-------/T :
     /T /T /T /1/|               /T/T /T /1/|               /T/T /T /1/|
    /--------/F / :             /--------/F /T:            /-------/F /T:
   /T /T /T /1/1/|             /T /T /T /1/1/|           /T /T /T /1/1/|
  :--------:F /F / :          :--------:F /F /* :        :--------:F /F /? :
  IE IE IE 1/1/1/             IE IE IE 1/ 1/ 1/          IE IE IE 1/ 1/ 1/
  :--------: / /              :--------: / /             :--------: / /
  I  IE I  1/1/               I  IE I  1/ 1/             I  IE I  1/ 1/
  :--------: /                :--------: /               :--------: R/
  I  I  I  1/                 I  I  I  1/                 I  I  I * 1/
  '--------'                  '--------'                  '--------'
```

Fig. 2D
```
Fig. 2D .--------.
       /T /T /? /|
      /--------/? :
     /T /T /T /1/|
    /--------/F / :
   /T /T /T /1/1/|
  :--------:F /F / :
  IE IE IE 1/1/1/
  :--------: / /
  I  IE I  1/1/
  :--------: R/
  I  I  I * 1/
  '--------'
```

Figure 2A is the same as Figure 1 except it is viewed from the left side and the (*) cube is to be used as the reference cube (1) during the translation.

STEP (1): Rotate the right side one Q turn toward the front as in Figure 2B.

STEP (2): Rotate the bottom layer one Q turn to the left as in Figure 2C.

STEP (3): Restore the top as in Figure 2D.

STEP (4): Again rotate the bottom layer one Q turn to the left as in Figure 2E. (This is the Q turn that churns the cubes in the bottom and/or middle layers).

STEP (5): Bring the right side down as in STEP (1).

STEP (6): Rotate the bottom layer two Q turns so it looks like Figure 2B again.

STEP (7): Restore the top so it looks like Figure 2A again.

If the fourth step was skipped over and STEP (6) rotated the bottom layer three Q turns instead, the CUBE after the seventh step would appear as it did before the first step, no changes would have occurred in the bottom and/or middle layers.

Fig. 2E
```
Fig. 2E .--------.
       /T /T /? /|
      /--------/? :
     /T /T /T /1/|
    /--------/F / :
   /T /T /T /1/1/|
  :--------:F /F / :
  IE IE IE 1/1/1/
  :--------: / /
  I  IE I  1/1/
  :--------: /
  IR I  I  1/
  '--------'
```

It takes seven steps for the CUBE to do this basic translation. Similarly, seven steps for the CUBE using either one of the other three reference cubes causes a different set of cubes to churn. NOTE: Since a translation on reference cube 1 does not involve the center or middle layers, the seven steps are accomplished by just seven twists of the CUBE. The other three take more than seven twists, e.g. it takes one twist to bring down reference cube 2 but two twists to move it out (one to rotate bottom and middle layers together, another twist to restore the bottom layer); it takes two twists to bring down reference cube 3 (one to bring down center and right slices together, another twist to restore the right side) but one twist to move it out; but for reference cube 4 it likewise takes two twists to bring it down and two twists to move it out. Therefore, the seven steps are accomplished by ten twists for reference cube 2, by eleven twists for reference cube 3, and by fourteen twists for reference cube 4. *IT WAS INDEED VERY ENCOURAGING THAT I WAS ABLE TO CHURN CUBES IN THE BOTTOM AND MIDDLE LAYERS AND THEN RESTORE THE TOP LAYER.*

When doing the translation on reference cube 1, as shown in Figure 2, I discovered that the four corner cubes did not change position, that three of the four corner cubes reoriented themselves (twisted) and that one remained fixed. I then proceeded to do translations on the other three reference cubes and discovered they had no effect on the corner cubes. After one of the many times that I messed up the top layer and had to start over again I discovered that the positions of the bottom four corner cubes had changed. So in order to get the bottom four-corner cubes in their correct position, I randomly kept messing up the top layer and reforming the CUBE as in Figure 1 until they were correct. Then after several translations using cube 1, the four corners popped up correct. Before retiring for the evening, I discovered that translations on cube 2 churned the four middle edge cubes but only one edge cube in the bottom layer. Monday evening was my greatest 'cubik' evening of the year. As soon as I got home from work I took copious notes on the movement of the cubes before and after each of the translations on the four reference cubes. By late evening I was able to correctly place (and orient) the bottom edge cubes by repeatedly using translations on cube 2. Then before retiring I was able to correctly place one or two of the middle edge cubes and orient them correctly by using translations on cube 4. In some instances the remaining four middle edge cubes were correctly placed but only two of them were correctly oriented. Wow, I only have two more cubes from making a solid CUBE. Tuesday evening my in laws came to our home and I was delighted to show them, especially my niece and nephew, my progress. I hardly worked on the CUBE that evening. Wednesday evening after band rehearsal and Thursday evening after tennis I put in a couple of hours each night trying to figure out how to flip around those last two cubes by trying several combinations of my translations. On Friday we had no plans and I, off and on, played with the CUBE now definitely getting frustrated since I was certain I could invert those last two middle edge cubes. I was ready to give the CUBE a heave (I didn't throw it yet) but I put it aside and watched some TV before calling it an evening. Before going to sleep it dawned on me that I could solve the CUBE this far without really needing the translation on cube 3. It is a popular translation. It only moves three edge cubes on one layer and during the translation two of the three cubes get flipped. I said to myself that tomorrow I'll have to try sending my two middle edge cubes down to the bottom layer, flip them around, and bring them back up and see what happens. On Saturday (one week after I purchased the CUBE) I finished my chores around noon, picked up the CUBE and lo and behold after about a half an hour it popped up whole. I ran into the kitchen to show my wife and my daughter was on the telephone and yelled over the phone 'o-my-gosh daddy just solved the Rubik's CUBE'. Since I had several algorithms down pat it was easy for me to remember the sequence of algorithms I used to flip (correctly orient) the last two middle edge cubes. That's how I solved the CUBE. It is a lengthy one but as I hope you'll see not too difficult to remember. From leafing through several other solutions I see in the bookstores, it is easy to arrive at the conclusion that mine is lengthy since it has no counter moves in the translations to further restrict the movement and orientation of the cubes to only one plane. That Saturday evening friends picked us up to see the Philadelphia Ballet. The man travels the world over as an engineering sales representative and when I told him about my luck he related what an accomplishment it was to solve the CUBE since he knew men all over the world that were literally being driven crazy by the CUBE.

Throughout the next week I took my CUBE to work keeping a low profile and during my lunch hour I would slowly walk eight blocks up side streets to a McDonald's playing with the CUBE all the way trying to improve my cubic finesse. Another time I was contemplating doing some patterns when on one occasion the CUBE was lying on my desk in its solid state and a fellow engineer, who never played with one before, gave it a couple of twists and up popped the 'X' pattern (I found out the next week that it is the 'The Pons Asinorum' configuration and I said to myself 'I don't have to worry about doing that pattern now'. Later that week after working hours I showed the CUBE to a fellow project engineer and our manager. The manager wasn't impressed and his only comment was to keep that menace at home since enough time is lost by personnel playing games with the computers and they don't need to bring CUBES to work adding to further lost time. Friday of that week was a tremendous 'cubik' day.

Lunchtime I meandered up to McDonald's and an engineer-musician colleague sat down at my table. After discussing some up and coming parades, I pulled the CUBE out of my raincoat pocket and told him the fun I found in solving it and demonstrated to him the ingenious mechanics that must be involved just to keep a corner cube in place as it is rotated in either of three planes. Why doesn't it just simply fall to the ground? My colleague marveled at that and definitely wanted to become a CUBE owner. Meanwhile, he told me that he saw an article about the CUBE on the front page of the Wall Street Journal. That was great, so I returned to work and found the paper and made a couple of Zerox copies. A fellow engineer-computer-buff saw the article and said he just received his Scientific American magazine and the CUBE made the front cover. He brought it to work on Monday, that was great, so lunchtime I purchased one on my way to McDonald's. Late Friday afternoon I saw my manager with the project engineer and went charging into his office with a CUBE and article in hand saying, 'Look, the CUBE made the front page of the Wall Street Journal'. Well, that did it, my managers attitude did two 'Q' turns. They gave it a couple of twists and marveled at the mechanics involved so I said, 'Wait a minute', and got the Scientific American and showed them the internal mechanism diagrams.

Out in the hallway after working hours I was showing my section manager the CUBE and the article when a young engineer friend, Bill, walked by and said, 'I have heard about the CUBE but have never seen one'. While Bill was giving it a couple of turns he said that he was going on a business trip next week to the west coast and would like to take a CUBE along. I could see that he was a potential cubist by his sincerity. I told him where to buy one and check that it has the little booklet. I gave him my last copy of the Wall Street Journal article and off he went.

The next week I spent solving the patterns that were in the Scientific American during my meandering to McDonald's, while in my car pool with an engineer neighbor, and at home. Also at home I took more notes on my translations while doing them either backwards or forwards using the four reference cubes shown in Figure 1 plus on translations using the two cubes to the left as reference cubes. It now took me almost twenty minutes to solve the CUBE and I wanted to improve my algorithms for positioning the bottom four-corner cubes, for correctly orienting the bottom edge cubes, and for finding an additional ending algorithm to handle the case when all the middle edge cubes came out in a worst-case condition. Using my algorithms in solving the various patterns didn't take much longer than when solving the solid CUBE pattern. The patterns looked so great that I bought my second CUBE, for me of course, since the first one was really my daughter's. Having two CUBES side by side with different patterns was the greatest. Anyway, by the end of the week my notes showed me an algorithm for positioning the bottom four-corner cubes instead of doing them randomly. Also doing a translation on cube 2 from the rear instead of from the front positioned and oriented immediately some of the bottom edge cubes. *(Currently inverting a face before doing a normal cube 2 translation accomplishes this.)* Still eluding me is an additional ending algorithm. I discovered that the translations using the two cubes to the left were redundant to translations using cube 1 and cube 2 in Figure 1.

Three weeks have now gone by since I bought my first CUBE. The next week Bill returned and he just had solved the CUBE. He works in an adjacent section so it is easy to get together on our coffee breaks. This was greater than great to have a fellow cubist nearby. His story is similar to mine and it took him about the same time with the same frustrations before he realized a solution. His time was around a total of eighteen hours of twisting the CUBE. He told me his solution involved solving all the edge cubes first paying no attention to the eight corner cubes. So he had to devise ending algorithms that maneuvered the corner cubes without disturbing the edge cubes. His dilemma was similar to mine in that after spending about ten hours on his solution he often got the CUBE to the point where there remained only two corner cubes to reorient. Like me, after spending another six hours, frustration started to creep in. He didn't give the CUBE a heave yet, either. On Friday evening he said that after trying some new translations to no avail, he thought, he put the CUBE aside for a while. After grabbing a snack he glanced over at the CUBE and had a feeling of surprise since the corner cubes looked different from what he thought they should have looked. Somehow he had flipped them. He immediately picked it up and after a short time was able to reconstruct his algorithm that flipped the two corner cubes and then in a short time it popped out in its pristine state. We thought about showing each other our solution but decided we would work on refining them at the present and agreed to share our solutions the next week.

During that same week when Bill returned from his business trip, I was trying to do the 'DOT' pattern starting from a scrambled CUBE. I was having difficulty since the last two middle edge cubes would always end up in opposite positions and I thought at that time since the CUBE has a zillion combinations that any pattern is possible. Well, I kept trying to interchange these two cubes by trying to devise additional algorithms. Well, what happen was, on Wednesday night I dropped my daughter off at her school and then went directly to band rehearsal and was very early. I was waiting in the car under a street lamp twisting my CUBE when all of a sudden the 'DOT' pattern popped out.

Wow, I couldn't believe it. The next day when I got to work I realized that my 'DOT' pattern was like the one in Scientific American and was not the one I started out to make. I wanted dots of opposite colors (same color arrangement as 'The Christman-cross' configuration). Then the thought hit me that the CUBE must have some mechanical limitations and some patterns are not possible, which happens to be true, and when I realized this fact, I felt relieved.

The next week Bill and I got together on our lunch hour and we took turns running through our solutions. Before doing them we both offered apologies to one another for fear something would go astray especially me since my ending for when the middle edge cubes came up in a worst-case scenario was still very cumbersome. I noticed his solution had more counter moves in his translations that allowed him to position more cubes in pairs and overall was shorter than mine but it looked trickier (more difficult to remember). Lucky for me my middle edge cubes came out all right and since my solution was so repetitious he recognized some translations that he also used. We talked for some time about our translations and how they led us to restoring it to a whole CUBE again. We have often pondered how much more difficult (and frustrating) it might have been if when we had gotten down to the last two or three cubes they all popped out in a solid state immediately, thereby, deceiving ourselves that we had solved the CUBE. However, as is obvious that was not the case and we had no difficulty solving the CUBE time and time again after it popped out solid that first time. At work there are some one hundred engineers on our floor (450 in the plant) and for the past year as far as we know not one other person would be able to sit down with Bill and I and run through their solutions with us. This is probably good testimony to the fact that the CUBE is referred to as a menace, a Hungarian horror, a hazard to your health, etc. It is just difficult to solve and not worth the time and effort to endlessly play with it until its mysteries are unraveled. The effort is certainly worth it but time is a definite obstacle. About ten engineers we know of received a CUBE throughout the year as a gift. A few of them solved it using Don Taylor's book, a couple did solve it but have great difficulty in repeating since they only have a very few algorithms (takes them hours); my engineer car pool friend solved the CUBE and is in this category. Several would like some help from Bill and I but are reluctant to get involved (several in this group are encouraging me to write this article) and finally the rest never took the CUBE out of the box. One evening during the week, I took notes on all the combinations I could think of using translations on cube 2 to send an edge cube from the middle layer to the bottom layer and then return it in order to find the additional ending algorithm that I was seeking. *(Currently my RM/0 operation does this.)* Well, it turned up and I was happy with my complete set of algorithms and haven't added to them since. Feeling happy I went out and bought myself my third CUBE. I said to myself that this one is for my brother so when I got home I used a razor blade to cut away the top cellophane and then scotch-taped the sides so that I could eventually put the CUBE back in the box and give it to him as a present with the tiny little booklet in tact. Before you give this article a heave instead of the CUBE, I better pause here in my 'cubik' story and start describing the solution. By the way, during the past year Bill and I kind of have a standing joke about the CUBE. On many occasions when I pass Bill in the hallway or see him at his desk I'll often say, 'Boy, where we lucky when the CUBE finally popped out all solid' and Bill always answers, 'Skill, skill! skill!!'

SOLUTION (Introduction) 1982

As indicated in the Figure 1 description, this solution is in five parts. The small letters t, f, r, e, b, k, are used to represent the six colors of the CUBE faces. (However they became capitol letters when using the ANDY font: T, F, R, E, B, K. for Figures 1 & 2 in this 2002 manuscript.) I now would like to use nine capitol letters to represent the nine different planes (slices) that the CUBE can be rotated in.

9 PLANES (Slices)

B = bottom,
M = middle (between top
T = top,   and bottom),
L = left,
R = right,
C = center,
Bk = back (rear),
F = front,
V = vertical (between
        back and front).

The symbols B, M, R, C, V, and Bk are used to label my algorithms or translations on reference cubes 1, 2, 3, and 4 or on slices. The table at the end of this article uses these symbols to outline all the translations needed to solve the CUBE. There are two identical tables so that one can be torn out. It is handy when making patterns from a scrambled CUBE.

SEVEN TRANSLATIONS/OPERATIONS (Table 1)

Translations On:	Name	Steps	Cyclic
Cube 1	R/B/1/1	8	3
Cube 2	R/V/1/1	7	5
Cube 3	C/B/1/2	7	3
Cube 4	C/V/1/2	8	2
Cube 1&2(R/0)	R/0//Bk/0	8 or 9	3 or 12
Cube 1&2(RR/0)	R/0//B/0//Bk/0	10	12
Cube 1&2(RM/0)	R/0//R/0//M/0	4	3
- - -	(ALGO1)	(45)	(2)

Hopefully, with the help of the CUBIK Algorithm Table and the CUBIK Flowcharts, it will be possible to waltz through a solution to the 'Rubik's CUBE'. Sheet 1 of the flowchart is an introduction to flowcharting. Sheet 2 assumes there is a CUBE in your midst and if it is scrambled it assumes you want to unscramble it either using any solution or using the CUBIK solution. If it is not scrambled, you could scramble it or make patterns from it. Sheet 3 starts the CUBIK solution and after selecting a top color (t) a FOUR STEP APPROACH sub-routine can be followed for completing the top layer.

* You are now about 150 twists away from making the CUBE whole. *

SOLUTION Part 1A
(Position Bottom Four Corners) (not to orient them)

Make sure your CUBE has all the top cubes oriented correctly as shown in Figure 1, and if not and you need help, a rather detailed description of how to solve one face of the CUBE using a FOUR STEP APPROACH is included after the flowcharts.

Rotate the bottom layer and see if by chance at least one of the bottom corner cubes can be positioned and oriented correctly as shown on the title page as reference cube 0 (zero). If none of the bottom four corner cubes line up, then doing a R/B/1/1 translation on either of the sides will line up at least one bottom corner cube when you rotate the bottom. The CUBIK Table shows a top view of what a R/B/1/1 translation will do. The front left corner cube remains unchanged after the translation and the other three corner-cubes twist. A sub-routine lists eight steps for doing this R/B/1/1 translation (CUBIK SUBROUTINES, Sheet 10). The first four steps were used in the example for Figure 2. The eighth-step is a correction step that returns the left corner (cube zero) to its proper corner by rotating the bottom layer two Q turns.

The R/B in the label identifies the reference cube looking at the top of the CUBE, so cube 1 in Figure 1 is the Right Bottom cube. The 1/1 in the label is a reminder to rotate the reference cube 1 Q turn in step 2 and to rotate it 1 Q turn in step 4.

If trying to do this R/B/1/1 translation you mess up the top, no problem, go back to the FOUR STEP APPROACH and restore the top layer.

Really, as implied by the flowchart, you don't have to do this above mentioned (optional) R/B/1/1 translation at this time but it is helpful (gives a good feeling) to have one corner cube correct (okay) to act as a reference point (cube zero) until these translations become familiar. In other words, it is possible to complete Part 1A without disturbing any orientation of the bottom left corner cube (cube zero).

Next thing to do according to the flowchart (Sheets 3 & 4), is to check to see if one, two, or all four bottom corner cubes happen to be in their correct corners (positioned correctly), remember they don't have to be oriented correctly, orienting the corner cubes is in Part 1B. The left front cube should be f, b, e; right front cube, f, b, r; left rear cube, k, b, e; right rear cube, k, b, r. I will now describe the operations that interchange two corner cubes. If only one corner is correct, rotate the bottom since at least two of the four corner cubes must be correct. As shown in Table 1 above, these operations are called R/0//Bk/0 (which is abbreviated as R/0) and R/0//B/0//Bk/0 (which is abbreviated as RR/0). As diagrammed in the CUBIK Table, the R/0 interchanges two adjacent corner cubes where as RR/0 interchanges two corner cubes diagonally.

Also, when doing the R/0 or RR/0 operations don't be concerned about the middle edge cubes moving nor the bottom edge cubes, let them move as they may, in other words, all we are concerned about is that after doing either of these operations the top layer of cubes (t) return to the way they are shown in Figure 1. Figure 3 illustrates the first four steps of the R/0 operation and the first six steps of the RR/0 operation. Sheet 10 of the sub-routine flowcharts lists the steps for these operations.

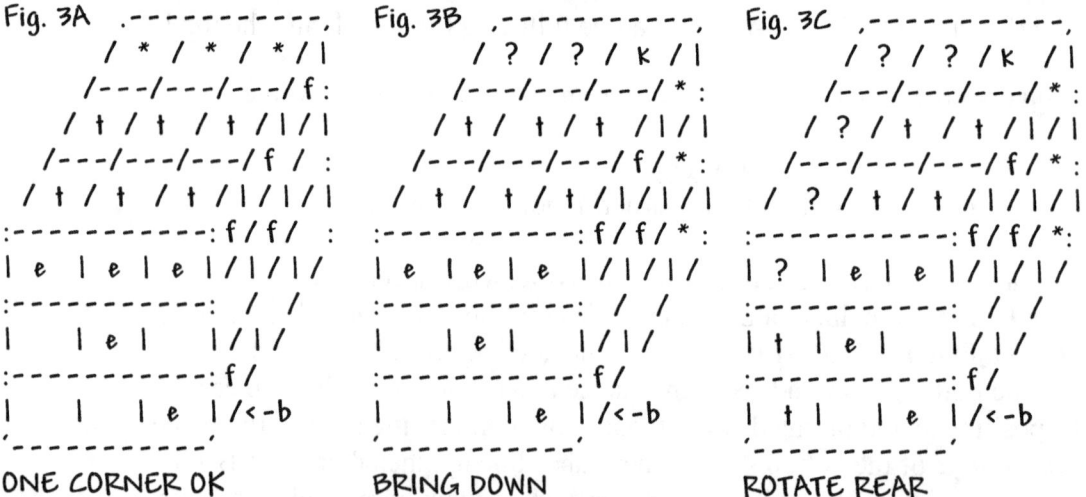

```
Fig. 3D  .-----------,
        / * / * / * /|
       /---/---/---/ f :
      / ? / + / + /|/|
     /---/---/---/ f/  :
    / ? / + / + /|/|/|
   :-----------: f/ f/  :
  | ? | e | e |/|/|/| 
 :-----------:  / /
 | + | e |  |/|/
:-----------: f/
| + |  |  | e |/<-b
'-----------'
BRING BACK
```

*(Three *'s are shown in Figure 3 since a R/O or a RR/O operation is now basically a slice operation.)*

R/O//BK/O

STEP 1: Bring down right side as in Fig. 3B,
STEP 2: Rotate rear 1 Q turn to the left as in Fig. 3C,
STEP 3: Bring back right side as in Fig. 3D,
STEP 4: Return rear 1 Q turn as in Fig. 3 E,
(#)STEPS 5,6,7&8 : Use the FOUR STEP APPROACH to recover the missing top cube (?): READY, BRING DOWN, MOVE IN, BRING BACK.
STEP 9: CORRECTION, rotate bottom layer 1 Q turn to the right which gets the bottom left front cube (f,b,e) back to its starting position as cube zero.

(#) READY: bottom layer 1Qt to the left; BRING DOWN: bring down back face; MOVE IN: bottom layer 1Qt to right; BRING BACK: restore top.

```
FIG. 3E  .-----------,
        / ? / * / * /|
       /---/---/---/ f :
      / + / + / + /|/|
     /---/---/---/ f/  :
    / + / + / + /|/|/|
   :-----------: f/ f/  :
  | e | e | e |/|/|/| 
 :-----------:  / /
 |  | e |  |/|/
:-----------: f/
|  |  | e |/<-b
'-----------'
ROTATE REAR
```

NOTE: The R/O operation is done on either one of the four faces (f,e,k,r) depending on those corner cubes you want to interchange. Figure 3A happens to be on the front face (f) and assumes the 'f,b,e' corner cube is correctly oriented, it really only needs to be positioned correctly. After STEP 9 is completed the two corner cubes at the rear will be interchanged as diagrammed in the CUBIK ALGORITHM TABLE. Likewise, Figure 3AA happens to be on the front face (f) but now the 'f,b,r' corner cube is oriented correctly, it really only needs to be positioned correctly. After STEP 10 of the RR/O operation is completed two diagonal cubes are interchanged as diagrammed in the CUBIK TABLE.

```
Fig. 3AA .-----------,            Fig. 3BB ,-----------,            Fig. 3CC ,-----------,
        / * / * / * /|                   / ? / ? / k /|                    / ? / ? / k /|
       /---/---/---/ f :                /---/---/---/ * :                 /---/---/---/ * :
      / + / + / + /|/|                 / + / + / + /|/|                  / + / + / + /|/|
     /---/---/---/ f/  :              /---/---/---/ f/ * :              /---/---/---/ f/ * :
    / + / + / + /|/|/|               / + / + / + /|/|/|                / + / + / + /|/|/|
   :-----------: f/ f/ f:-#         :-----------: f/ f/ * :           :-----------: f/ f/ r :-#
  | e | e | e |/|/|/|              | e | e | e |/|/|/|                | e | e | e |/|/|/|
 :-----------:  / /               :-----------:  / /                 :-----------:  / /
 |  | e |  |/|/                   |  | e |  |/|/                      |  | e |  |/|/
:-----------:  /                 :-----------:  /                    :-----------: r /
|  |  |  |/  #=frb               |  |  |  |/                         |  |  | * |/  #=frb
'-----------'                    '-----------'                       '-----------'
ONE CORNER OK                    BRING DOWN                          ROTATE BOTTOM
```

```
Fig. 3DD.----------,        Fig. 3EE ,----------,        Fig. 3FF ,----------,
     / ? / ? / K /|             / ? / ? / K /|               / * / * / * /|
    /---/---/---/ * :          /---/---/---/ * :            /---/---/---/ f :
   / ? / t / t /|/|           / ? / t / t /|/|             / ? / t / t /|/|
  /---/---/---/ f / * :      /---/---/---/ f / * :        /---/---/---/ f /  :
 / ? / t / t /|/|/|         / ? / t / t /|/|/|           / ? / t / t /|/|/|
:----------: f / f / r :   :----------: f / f / * :     :----------: f / f / f :-#
| ? | e | e|/|/|/        | ? | e | e|/|/|/          | ? | e | e|/|/|/
:----------: / /        :----------: / /            :----------: / /
| t | e |   |/|/         | t | e |   |/|/            | t | e |   |/|/
:----------: r /        :----------: t /            :----------: t /
| t |   | * |/           | ? |   | K |/              | ? |   | K |/    #=frb
'----------'            '----------'                '----------'

ROTATE REAR              RETURN BOTTOM               BRING BACK
Fig. 3GG.----------,
     / * / * / * /|
    /---/---/---/ f :         R/0//B/0//BK/0
   / ? / t / t /|/|        STEP 1: Bring down right side as in Figure 3BB,
  /---/---/---/ f /  :      STEP 2: Rotate bottom 1 Q turn to the left as in Figure 3CC,
 / ? / t / t /|/|/|         STEP 3: Rotate rear 1 Q turn to the left as in Figure 3DD,
:----------: f / e / f :-#  STEP 4: Return bottom as in Figure 3EE,
| ? | e | e|/|/|/          STEP 5: Bring back right side as in Figure 3FF,
:----------: t / /         STEP 6: Get READY two top cubes as in Figure 3GG
|   | K | K|/|/                       (middle layer 1 Q turn to the right),
:----------: t /           STEPS 7,8&9: BRING DOWN, MOVE IN, and BRING BACK
|   |   | K |/   #=frb                  to restore the top layer,
'----------:              STEP 10: CORRECTION, rotate bottom layer 1 Q turn to the
READY                                 right in order to get bottom right front cube
                                       to staring position, 'f,r,b' in Figure 3AA.
```

One thing that amazes me about the CUBE is that all the translations or operations are cyclic as mentioned before. The cyclic column in Table I indicates that if you repeat the R/0 (9 stepper) operation exactly twelve times starting with a solid CUBE it would churn the little cublets during 108 twists of the CUBE and then it returns to a solid CUBE again after the 108th twist. If you ever try this, you will notice that after each R/0 operation that these operations only churn cubes in the bottom layer. After repeating the R/0 operation only six times the corner cubes are all okay but two of the edge cubes are not oriented. Similarly, after repeating the RR/0 operation six times the corner cubes are okay but two of the bottom edge cubes are not oriented. However, the RR/0 operation churns cubes in both the bottom and middle layers. If step nine is eliminated, the R/0 operation (now an eight stepper) is now cyclic in three cycles instead of twelve but also after the first and second cycles the reference cube zero is not in its starting position.

SOLUTION Part 1B
(Orient bottom four corners)

Now that Part 1A is done and the bottom four corner cubes are in their correct position, doing one, two, or maybe up to four R/B/1/1 translations correctly orients them. The R/B/1/1 translation does not disturb the bottom left front corner cube and twists the other three corner cubes as shown in the CUBIK TABLE. It also moves around the bottom edge cubes, forget about them, they get positioned and oriented in the next part, Part 2. This R/B/1/1 translation does not disturb the cubes in the top layer nor in the middle layer, in other words when you finish with the translation all the cubes in the top and middle layers return to the same position they had before doing the translation. To recall the R/B/1/1 translation sequence that takes eight steps, see Sheet 10 of the CUBIK SUBROUTINES.

```
Fig. 4A  .-----------,      Fig. 4B  ,-----------,      Fig. 4C  ,-----------,
        / + / + / + / |             / + / + / + / |             / + / + / + / |
       /---/---/---/ r :           /---/---/---/ r :           /---/---/---/ r :
      / + / + / + /\/ |           / + / + / + /\/ |           / + / + / + /\/ |
     /---/---/---/ r /  :        /---/---/---/ r /  :        /---/---/---/ r /  :
    / + / + / * /\/\/ |         / + / + / * /\/\/ |         / + / + / + /\/\/\/ |
   :-----------: r / r / k:    :-----------: r / r / b:    :-----------: r / r / r:
   | f | f | f | f |/|/| /     | f | f | f | f |/|/| /     | f | f | f | f |/|/| /
   :-----------:  / /         :-----------:  / /         :-----------:  / /
   |   | f |   |/|/          |   | f |   |/|/          |   | f |   |/|/
   :-----------: b /         :-----------: f /         :-----------: r /
   | f |   | r |/            | f |   | b |/            | f |   | f |/
   '-----------'             '-----------'             '-----------'
      IN  SYNC                  IN  SYNC                 CORNERS DONE
```

In order to get the corner cubes to pop out OK, they first must be 'in-sync'. Figure 4A and Figure 4 B show the bottom left front corner cube is OK and the other three corners cubes are 'in-sync'. The (*) cube is just a reminder to do a R/B/1/1 translation. When the corners get 'in-sync' they will be like Figure 4A, otherwise like Figure 4B. For Figure 4A the colors on the side of the bottom layer going left to right are f/ r,b / k,b / e,b / e. Now doing an R/B/1/1 translation on the front face (f) of Figure 4A results into Figure 4B where the colors going left to right are f/ b.f / b,r / b.e / e. Notice the bottom (b) color is now changed to the left. Remember that the bottom color (b) is the color of the bottom face of the CUBE. Next doing an R/B/1/1 translation on the front face (f) of Figure 4B results into Figure 4C which is what we want, all the corner cubes are OK now, so after the CUBE looks like Figure 4C, go on to the next part, Part 2.

Check to see if your bottom corners are already 'in-sync', like Figures 4A or 4B. If so, you have got it made, one or two R/B/1/1 translations and you are done. If not so, I will describe how you get them 'in-sync'. Meanwhile, in doing the R/B/1/1 translations you have messed up the top by accident, immediately restore it using the FOUR STEP APPROACH, and there is a good chance you have not disturb the position of the bottom four corner cubes, otherwise, if they are disturbed you must go back to Part 1A and readjust them.

Now to get them 'in-sync', check the corners and see if one is OK, if two are OK, or none are OK. Until they are 'in-sync', the corners will be in one of these three categories. If one is OK, the other three must be 'in-sync', if they are not 'in-sync', the CUBE must have been tampered with. If none are OK do a R/B/1/1 translation on any of the four faces ('f', 'r', 'k', 'e') arbitrarily, then two or one of the corners will be OK. If one is OK they must be 'in-sync'. However, when two corners are OK, doing a R/B/1/1 translation in accordance with the CUBIC FLOWCHART and the CUBIC TABLE results in one corner becoming OK and the other three 'in-sync'. To get them 'in-sync' with one or two R/B/1/1 translations does take some practice.

Here again, in order to get the bottom four corners 'in-sync', my original manuscript was not too precise and just after producing the "Let's Do It" exercise in the pattern section I decided to produce an OUT-OF-SYNC subroutine in the flowchart section which in conjunction with the CUBIK TABLE is precise.

Often when I can't get them 'in-sync' I check back and find out that I had goofed, I still didn't position the corner cubes correctly, so recheck that all the corner cubes are in their correct corners. Last summer my oldest daughter's boyfriend disassembled one of our CUBES and it was interesting to see the guts of the CUBE and inspect the different shapes of the little cublets. He then assembled it and told me that he had put the pieces back at random. Interesting, I thought, I would try to make it one solid color again. Well, after doing the top layer and positioning the bottom four-corner cubes, I started to do R/B/1/1 translations in order to orient them correctly. Not only couldn't I get them 'in-sync', the pattern of the bottom four bottom corners looked strange. So when I reasoned where they should be 'in-sync' and they weren't (it appeared as if two cubes looked correct), I took a screwdriver and slightly lifted the other one and twisted it. Now they were 'in-sync'. So if your CUBE has had its corner cubes tampered with, its trying to do this Part 1B that will discover such tampering.

Well the gist of this 'in-sync' business is that after doing Part 1A, which positions the bottom corner cubes, usually when you examine them they are very seldom 'in-sync' and either two of the corners are OK or none are OK. So if at random you pick a corner and start doing R/B/1/1 translations you will find either one corner becomes OK (other three corners are 'in-sync' like Figures 4A and 4B) or the corners keep alternating between none are OK to two are OK as you continue doing R/B/1/1 translations. The gist is to break up this alternating situation and get them 'in-sync' so that once 'in-sync' two more R/B/1/1 translations will orient all the bottom corner cubes as shown by Figures 4A, 4B, and 4C.

NOTES:

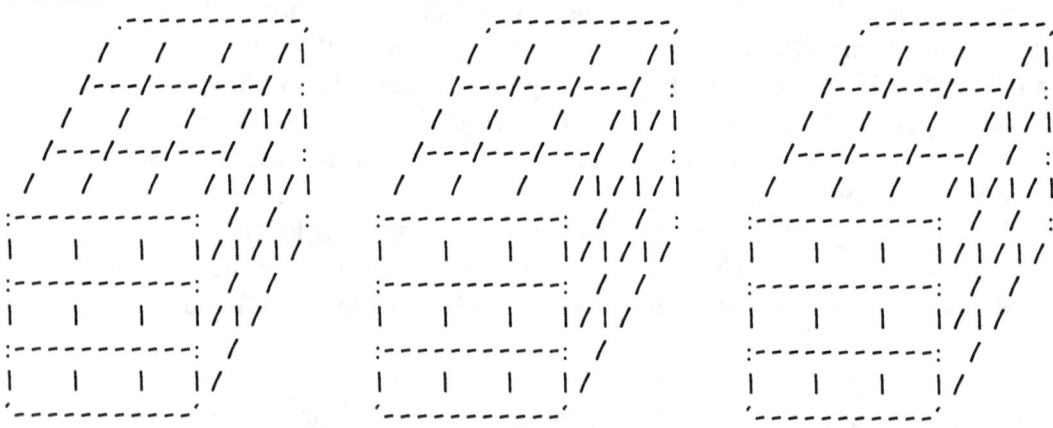

The OUT-OF-SYNC subroutine handles the situation when two of the four corners are oriented correctly and the CUBIK TABLE illustrates how to get them 'in-sync'. When none are OK, the diagrams below show eight outcomes when randomly doing a R/B/1/1 translation on one of the four faces ('f', 'r', 'k', 'e'): four outcomes put them 'in-sync', two outcomes make two corners OK that are adjacent, and two outcomes make two corners OK that are diagonal. (THIS IS OPTIONAL AND IS NOT PART OF THE CUBIK FLOWCHART, FOR ACADEMIC INTEREST ONLY.)

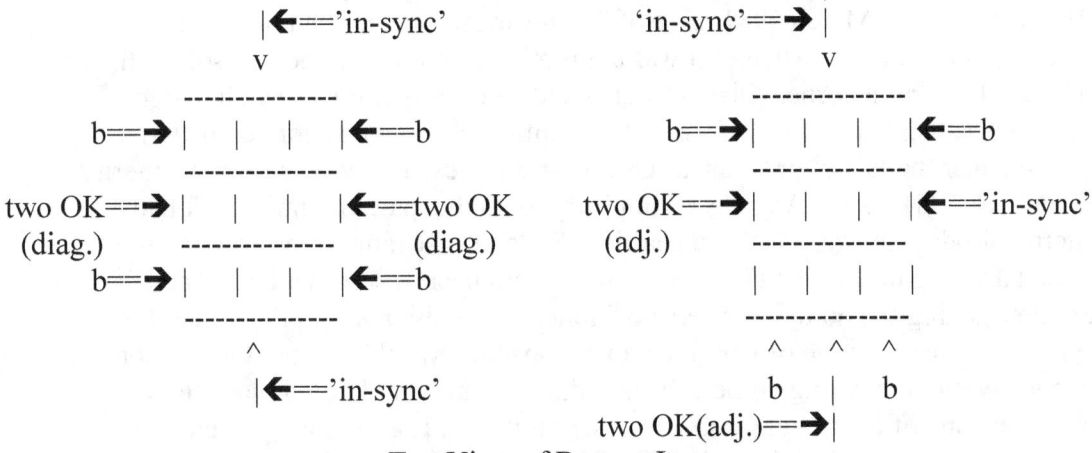

Top View of Bottom Layer

(Two cases when all the bottom four-corner cubes have their bottom color to the side.)

The CUBIK TABLE diagrams the one case for holding the CUBE when doing a R/B/1/1 translation on the face facing you when the two OK corner cubes are on a diagonal (OOS3). Two cases are shown when the two OK corners are adjacent (OOS1 & OOS2). In all three cases, one R/B/1/1 translation gets them 'in-sync'.

The R/B/1/1 translation is cyclic in three cycles. So if you have a solid CUBE and do three consecutive R/B/1/1 translations on the same face (a total of twenty-four twists) it will turn up one solid color again. After each translation notice that the bottom four corner cubes follow the same pattern as in Figures 4A, 4B, and 4C. This R/B/1/1 translation does not disturb the bottom left front corner cube (cube zero), the bottom left edge cube, nor the bottom center cube. It does move around the other three bottom edge cubes.

Part 1 Summary

What can I say, by now you should be very familiar with the two operations and one translation involving the bottom four corners (R/O, RR/0, R/B/1/1). Playing with these operations and translations finishes the bottom corner cubes. In Part 2 we use a translation on reference cube 2 (R/V/1/1) in order to complete the bottom layer, however, we don't do as much playing with it as we did for Part 1. Since Part 1 is now done all eight corner cubes are OK, the remaining translations only translate edge cubes so from now on if you make a mistake it usually only involves the edge cubes so if you immediately correct those edge cubes that were disturb it is very unlikely that you will disturb any of the corner cubes. If you do disturb the corner cubes, too bad, go back to Part 1.

Since getting the cubes 'in-sync' appears a little tricky, I would like to digress a little bit here before going on to Part 2. When doing patterns from a scrambled CUBE, I use the FOUR STEP APPROACH to arrange the pattern on the top layer and the bottom layer usually becomes some image of the top. With a pattern it now becomes trickier to get the bottom corners 'in-sync', but it is an interesting challenge. With regards to getting the bottom corners 'in-sync', consider the BIT MASON pattern (pattern P30A or P30B in the patterns from a scrambled CUBE section, page P5). This is the pattern that is on the front cover of the March 1981 issue of Scientific American. I admit that the first time I saw it I didn't know whether it was a gimmick or not (for I had just solved the CUBE). The top front right corner cube is twisted and since corner cubes can be oriented three different ways, there are two versions of this configuration. I categorized patterns as belonging to either the 'slice' or 'Mason' groups. I got these two words from a pattern that is in the March 1981 issue. Anyway, I didn't try to do this pattern until after I had my dot pattern episode that was mentioned earlier. So later that same week I started this pattern and had a strong hunch that the cube in the bottom back left corner had to be twisted. So after getting the bottom corners positioned correctly, I would think that they were 'in-sync' and after a couple of R/B/1/1 translations they would pop out OK except the twisted cube was in the wrong corner. It soon dawn on me that I could place the twisted cube in anyone of the bottom four corners. All this is, is a long way to tell you that after you do a solid CUBE, doing this BIT MASON arrangement from a scrambled CUBE is a challenge to get the bottom corners 'in-sync' and then have them pop out with the twisted corner diagonally opposite to the twisted corner seen on the front cover. Usually a lot of patterns are like this, any pattern that appears over three faces of the CUBE such as 't', 'f', and 'r' faces, would also appear as a direct image over the 'b', 'k', and 'e' faces. The 'slice' and 'Mason' patterns are interesting groups. Several of them are shown in the pattern section appearing after the flowcharts. One other thing, the rest of my translations do not require cubes being 'in-sync' or anything like that.

* You are now about 95 twists away from making the CUBE whole. *

<u>SOLUTION Part 2</u>
(Position and Orient Bottom Four Edge Cubes)

Make sure your CUBE has all the top cubes and bottom corner cubes oriented correctly as shown in Figure 5. If your CUBE looks like Figure 5, look at the bottom face and sides and see if by chance one or more of the four bottom edge cubes are already positioned and oriented correctly. If all four bottom edge cubes are OK (hardly ever happens) you've done Part 2, go on to Part 3 and your CUBE will look like Figure 9. If one, two, or three are OK, they won't be disturb, all that remains is to position and orient the other(s).

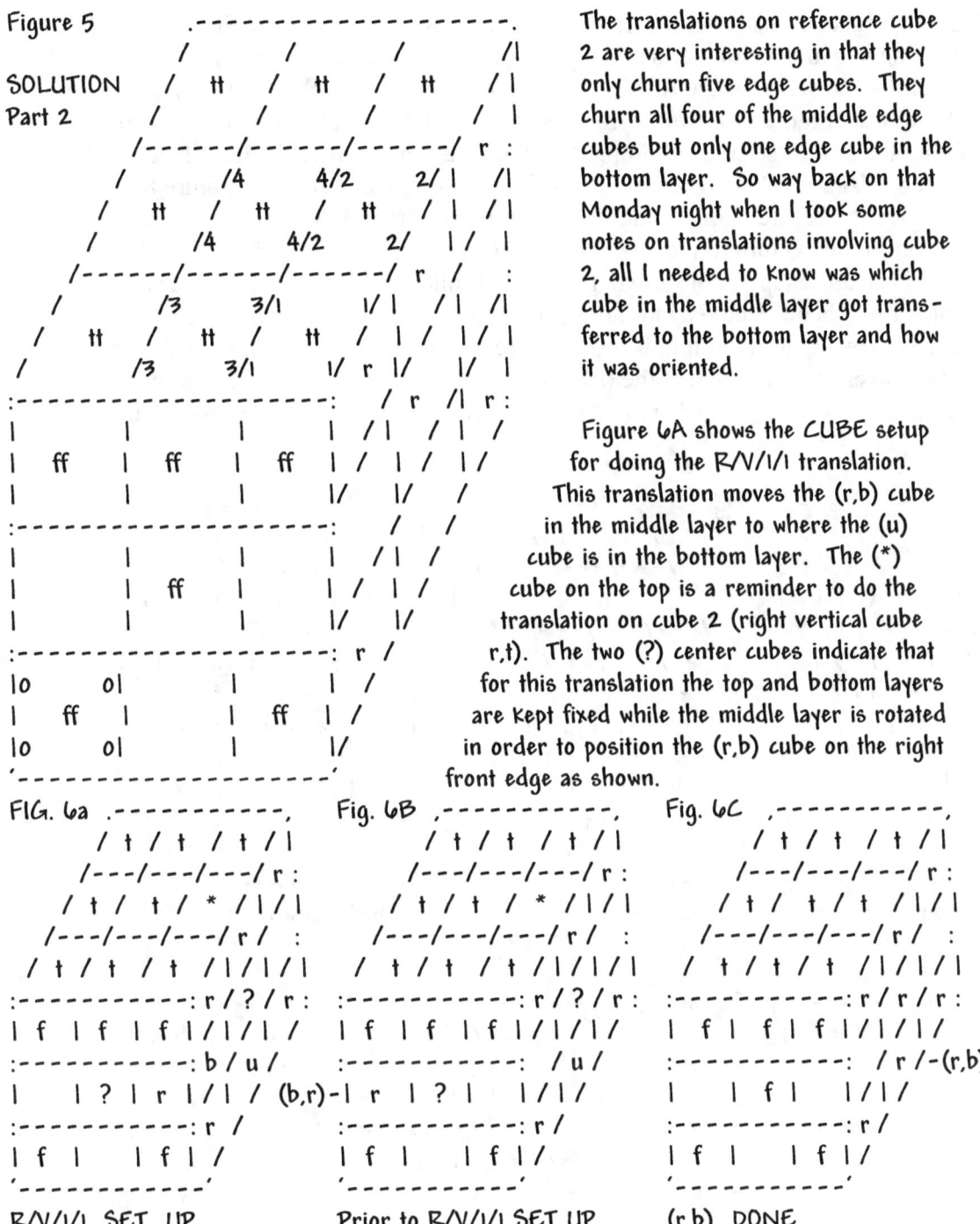

Figure 5
SOLUTION
Part 2

FIG. 6a — R/V/1/1 SET UP

Fig. 6B — Prior to R/V/1/1 SET UP

Fig. 6C — (r.b) DONE

The translations on reference cube 2 are very interesting in that they only churn five edge cubes. They churn all four of the middle edge cubes but only one edge cube in the bottom layer. So way back on that Monday night when I took some notes on translations involving cube 2, all I needed to know was which cube in the middle layer got transferred to the bottom layer and how it was oriented.

Figure 6A shows the CUBE setup for doing the R/V/1/1 translation. This translation moves the (r,b) cube in the middle layer to where the (u) cube is in the bottom layer. The (*) cube on the top is a reminder to do the translation on cube 2 (right vertical cube r,t). The two (?) center cubes indicate that for this translation the top and bottom layers are kept fixed while the middle layer is rotated in order to position the (r,b) cube on the right front edge as shown.

This R/V/1/1 translation correctly orients the (r,b) cube in the bottom layer. If this (r,b) cube happens to be reversed, colors are (b,r) instead, this translation positions the cube in the bottom layer correctly but it does not orient it correctly. Figure 6B shows the CUBE set up prior to doing the R/V/1/1 translation when the cube is reversed. (The 'b' face of the 'b.r' cube is not shown since it is on the left face.) Now rotating the front two Q turns (inverts it) puts the (b,r) cube in a (r.b) position so a R/V/1/1 translation places it correctly in the bottom layer. This translation for positioning and orienting the bottom four edge cubes is diagrammed in the CUBIK TABLE. Before describing a strategy for using this translation, let's go over the seven steps used in doing this translation. The seven steps for the R/V/1/1 translation are listed on Sheet 11 of the CUBIK subroutines. Figure 7 shows the first three steps of doing a R/V/1/1 translation. Figure 8 shows the first three steps of doing a R/V/1/1 translation after the front face is inverted. Figures 7A and 8A show the right side being brought down but in Figure 8A the front was inverted before being brought down. Figures 7B and 8B show their middle layers being rotated one Q turn to the left. Figures 7C and 8C show their right sides being brought back up. Step 4, not shown, rotates their middle layers another Q turn to the left (churns the cubes). Step 5 again brings down the right side. Step 6 moves the reference cube (*) two Q turns so it looks like Figures 7A or 8A. Step 7 recovers the top right side. (Re-invert the front face if the (r,b) cube was reversed.) Note that Figure 6 shows the front face towards you where as Figures 7 and 8 show the left face toward the reader.

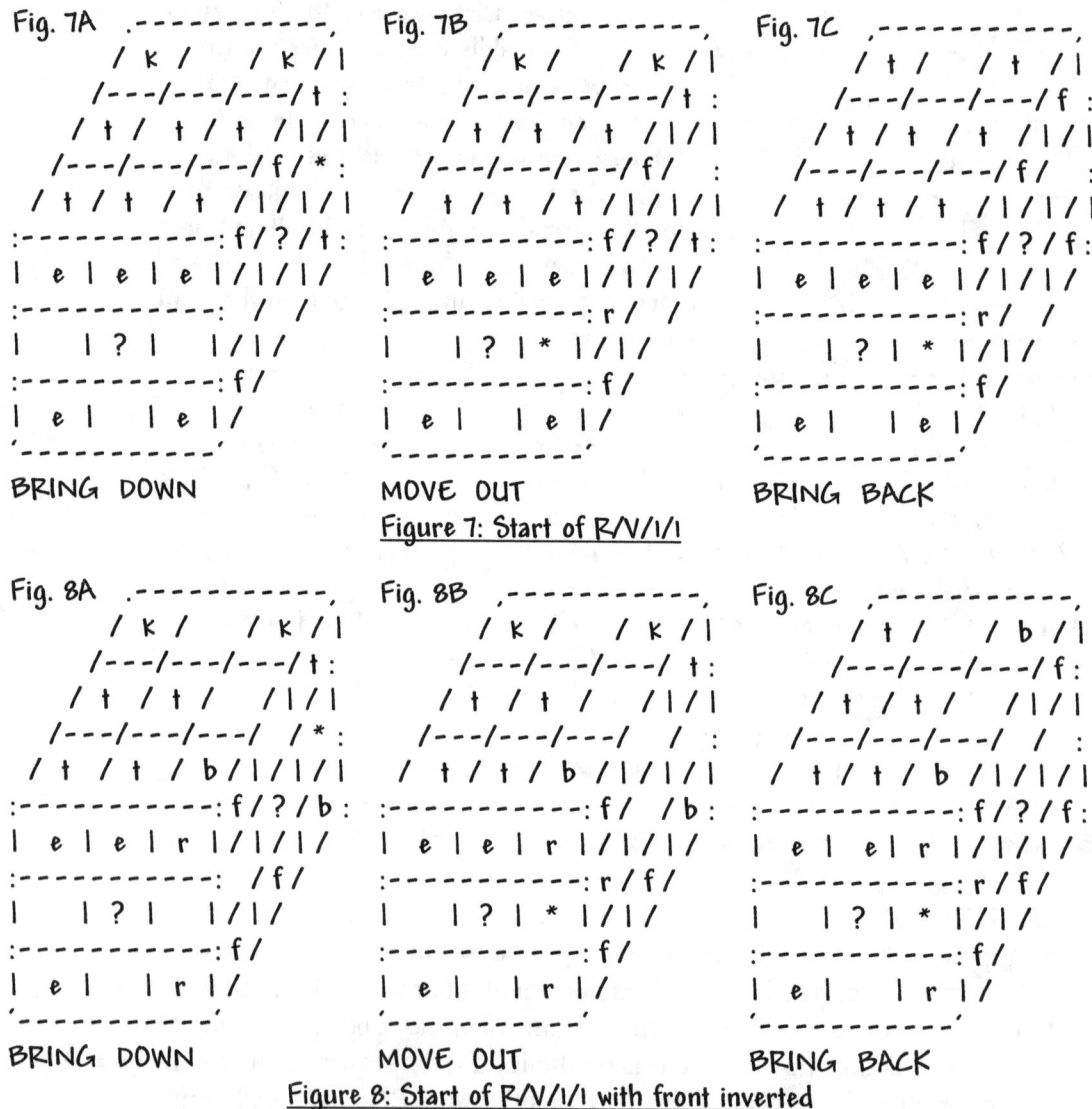

Figure 7: Start of R/V/1/1

Figure 8: Start of R/V/1/1 with front inverted

Let's assume you just completed the R/V/1/1 translation which correctly moved the (r,b) cube to the bottom layer, then it is important to rotate the middle layer back to its correct position as shown in Figure 6C before deciding which cube to move to the bottom layer next. If you don't do this, you will find out especially when doing patterns that it is very easy to put the bottom layer edge cube in the wrong position. If this happens, don't worry about it, it just means that you have to do the translation over again. Now to discuss the strategy in moving edge cubes one-by-one to the bottom layer. First let me say that when doing these translations you will eventually increase your speed and mistakes result. Often the case is that you dip the right side instead of restoring it and that can mess up the top layer. If this happens, try to restore the top as soon as possible using the FOUR STEP APPROACH and it is very unlikely that you will disturb any of the eight corner cubes, if you do disturb them, back to Part 1.

As shown by the flowchart Sheet 6 any bottom edge cube(s) in the middle layer can be immediately moved to the bottom layer. If there are no more bottom edge cubes in the middle layer, then some of the edge cubes in the bottom layer must be wrong, therefore, place the incorrect edge cube in the 'u' position and use the R/V/1/1 translation to bump it back from the bottom layer to the middle layer, then it can be properly transferred to the bottom. Finishing the bottom four-edge cubes completes the bottom layer.

Part 2 Summary

By now you should be very familiar with the translation on cube 2 (R/V/1/1). Using this translation finishes the bottom edge cubes; therefore, there are only four middle edge cubes that remain to be placed. Basically, so far after using the FOUR STEP APPROACH to finish the top layer, three basic translations are required to finish the bottom layer, R/0 or RR/0 (which are now slice operations), R/B/1/1, and R/V/1/1. In order to finish the CUBE, two additional translations are required and they are C/V/1/2 and C/B/1/2 and one of the main reasons for me to write this story is that this algorithm approach basically uses four translations (or algorithms), namely: R/B/1/1, R/V/1/1, C/V/1/2, and C/B/1/2, that are highly similar and they require no notations since once described they become repetitive, used over and over again. Also, three slice operations, R/0, RR/0, and RM/0 are required.

Originally I used a L/V/2/1 translation, but done from the rear of the CUBE on cube 2, to transfer a 'reversed' edge cube to the bottom layer. Also in order to rotate three middle edge cubes, I used this L/V/2/1 translation to send a middle edge cube to the bottom layer followed by a R/V/1/1 translation to bump it back up (this rotation is now done by the current RM/0 slice operation). My original R/0 and RR/0 operations affected the bottom corner cubes, R/0 rotated three of the corners, RR/0 interchanged two adjacent corners. They both operated on reference cube 1, dipping it down, moving it out, and dipping further the right side that disturb the top layer, etc. Now two slice operations interchange corner cubes. All of this to say that my original solution focused on combinations of my seven step sequences, namely: R/B/1/1, R/V/1/1, L/V/2/1, C/V/1/2, and C/B/1/2 where R/0 and RR/0 were deviations on the R/B/1/1 sequential steps.

Consequently, as a result of these seven 'current' algorithmic operations, I was able to construct a flowchart for solving the 'RUBIK's CUBE' which appears at the end of this story. These seven operations (algorithms) are tabulated and diagrammed in a

CUBIK TABLE appearing at the end of the story.

I hope in completing Part 2 you have become very familiar with the translation using cube 2 as the reference cube since this translation is used many times in Part 5 which is part of my ending algorithm which I will call ALGO#1.

* You are now about 70 twists away from making the CUBE whole. *

<u>SOLUTION Part 3</u>
(Positions at least one middle edge cube)

Make sure your CUBE has all the top layer cubes and the bottom layer cubes oriented correctly as shown in Figure 9. If your CUBE is like Figure 9, check the middle edge cubes and see if by chance one or more of them are already correctly positioned. If all four middle edge cubes are OK (hardly ever happens), the CUBE should be a solid, YOU'RE FINISHED. If one is OK, go to Part 4, if two are OK, go to Part 5, and if none are OK, stay here.

Figure 9

SOLUTION
Part 3

```
                .---------------------.
               /      /      /      /|
              /  tt  /  tt  /  tt  / |
             /      /      /      /  |
            /------/------/------/ r :
           /    /4     4/2    2/ |  /|
          /  tt  /  tt  /  tt  / | / |
         /    /4     4/2    2/  |/  |
        /------/------/------/ r /   :
       /    /3     3/1    1/ |  /|  /|
      /  tt  /  tt  /  tt  / | / | / |
     /    /3     3/1    1/  1/  1/   |
    :--------------------: r / r /| r :
    |      |      |      | /|  / | /
    |  ff  |  ff  |  ff  | / | /  |/
    |      |      |      |/  |/   /
    :--------------------:   / r /
    |      |      |      | / | /
    |      |  ff  |      |/  | /
    |      |      |      |/  |/
    :--------------------: r /
    |o    o|      |      | /
    |  ff  |  ff  |  ff  | /
    |o    o|      |      |/
    '--------------------'
```

The translation on cube 4 only disturbs the four middle edge cubes. It crisscrosses them as shown in the CUBIK TABLE. The translation on cube 4 is labeled C/V/1/2. The eight steps for doing a C/V/1/2 translation are listed on Sheet 11 of the CUBIK subroutines. Figure 10 shows the first three steps of doing a C/V/1/2 translation. Figure 10A shows bringing down the center slice one Q turn. Figure 10B shows the middle layer being rotated one Q turn to the left. Figure 10C shows bringing back the center slice. Step 4, not shown, rotates the middle layer an additional two Q turns to the left. Step 5 again brings down the center slice. Step 6 moves the reference cube (*) one Q turn so it looks like Figure 10A. Step 7 recovers the top. Step 8 realigns the center cubes by rotating the middle layer two Q turns. Doing a C/V/1/2 translation on either of the four faces (f, e, k, or r) gets the same result by crisscrossing the four middle edge cubes. Figure 10 happens to be on the front face 'f'.

Sheet 7 of the CUBIK flowchart shows the strategy in trying to position at least one middle edge cube. Carefully look at the middle edge cubes and see if swapping (interchanging) and two of them that are diagonally across from one another will correctly position one of them. If so, doing a C/V/1/2 translation on any face will at least make one middle edge cube correctly positioned (doesn't have to be oriented) and you can go on to Part 4. Instead of two of them being diagonally across, if one can be made correct by swapping them straight across (interchanging them either along the right face or the left face), rotating the front face two Q turns (invert it) before doing the C/V/1/2 translation will make at least one middle edge cube correctly positioned and you can go on to Part 4.

```
Fig. 10A .-----------,        Fig. 10B ,-----------,       Fig. 10C ,-----------,
      / t / t / t /|               / t / t / t /|              / t / t / t /|
     /---/---/---/ f:            /---/---/---/ f:            /---/---/---/ f:
    / k / k / k /|/|             / k / k / k /|/|           / t / r / t /|/|
   /---/---/---/ t/  :          /---/---/---/ t/  :        /---/---/---/ f/  :
  / t / t / t /|/|/|           / t / t / t /|/|/|         / t / t / t /|/|/|
  :-----------: f/ * /f:        :-----------:f/ r /f:      :-----------:f/ f /f:
  | e | e | e |/|/|/           | e | e | e |/|/|/         | e | e | e |/|/|/
  :-----------:  /t/            :-----------:  /t/         :-----------:  /f/
  |   | e |   |/|/             |   | * |   |/|/           |   | * |   |/|/
  :-----------:f/               :-----------:f/            :-----------:f/
  | e | e | e |/                | e | e | e |/             | e | e | e |/
  '-----------'                 '-----------'              '-----------'
    BRING  DOWN                    MOVE OUT                   BRING  BACK
```

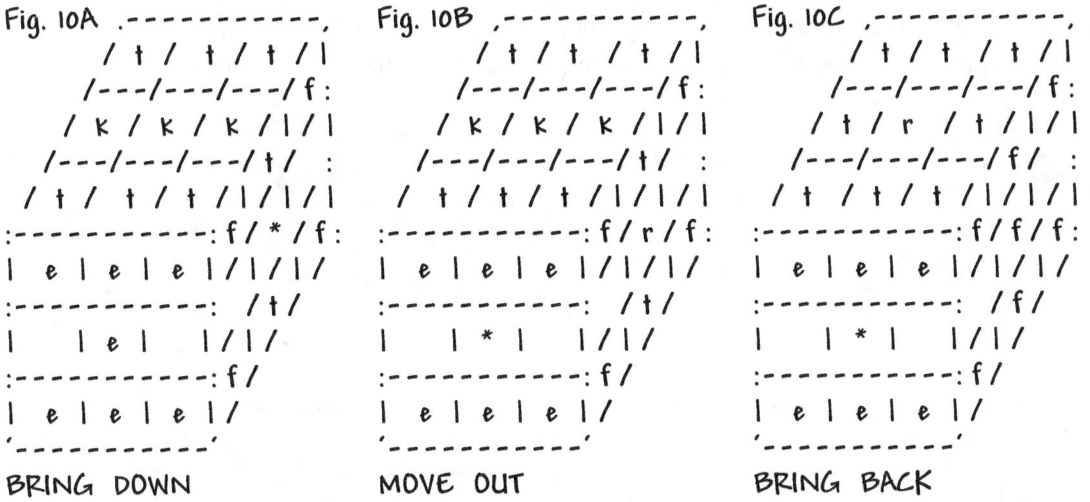

<u>Figure 10: Start of C/V/1/2</u>

     The C/V/1/2 translation is cyclic in two cycles. So if you have a CUBE in its pristine state and do two consecutive C/V/1/2 translations (a total of sixteen twists) it will turn up in one solid color again. As indicated before, this translation only affects the four middle edge cubes.

<u>Part 3 Summary</u>
     Part 3 is the only time the C/V/1/2 translation is used in the Cubik solution. By carefully looking at the middle edge cubes, only one C/V/1/2 translation is necessary in order to place correctly one of these four edge cubes on its correct edge. Remember, if before doing the C/V/1/2 translation you rotate the front face two Q turns (invert it), the four middle edge cubes are interchanged along the sides of the CUBE instead of diagonally. Also, always do the CORRECTION step (Step 8) before readjusting the front face, in other words, move the middle layer two Q turns (Step 8), and then move the front face two Q turns. It is also important to remember that for the C/V/1/2 and the C/B/1/2 translations, Step 4 rotates the middle or bottom layer two Q turns, whereas for the R/B/1/1 and the R/V/1/1 operations Step 4 rotates the bottom or middle layer only one Q turn.

    * * You are now about 60 twists away from making the CUBE whole. * *

23

SOLUTION Part 4
(Positions the second middle edge cube.)

Make sure your CUBE has all the top layer cubes plus all the bottom layer cubes OK and one of the four middle edge cubes positioned correctly but perhaps not oriented correctly as shown in Figure 11. While doing Part 4 and you find out that you are unable to position the four middle edge cubes correctly, then your CUBE has been tampered with. Get a screwdriver, disassemble the CUBE and interchange them so that they are correct.

NOTES:

```
Figure 11            .------------------.
                    /     /     /     /|
SOLUTION          /  tt  /  tt  /  tt / |
Part 4          /     /     /     /  | |
             /------/------/------/ r  :
           /      /4     4/2    2/ |   /|
         /   tt  /  tt  /  tt  / | / |
       /      /4     4/2    2/   | /  |
      /------/------/------/ r  /    :
      /    /3    3/1     1/ |   /|   /|
    /  tt  /  tt  /  tt  / |  / |  / |
   /    /3    3/1     1/ r |/  |/  |
  :--------------------------:  / r /| r  :
  |      |      |      |  /|  / | /
  |  ff  |  ff  |  ff  | / | / | /
  |      |      |      |/  |/  /
  :--------------------------: r  /  r /
  |      |      |      |  /| /
  |      |  ff  |  ff  | / | /
  |      |      |      |/  |/
  :--------------------------: r  /
  |o    o|      |      |  /
  |  ff  |  ff  |  ff  | /
  |o    o|      |      |/
  '--------------------'
```

One or more RM/0 translations position the second middle edge cube (which automatically correctly positions the remaining two). As shown by the CUBIK TABLE, RM/0 keeps one of the edge cubes fixed but rotates the other three middle edge cubes. The sequence for doing a R/0//R/0//R/M sequence (abbreviated as a RM/0 operation) is listed on Sheet 12 of the CUBIK subroutines. Figure 12 diagrams the five steps in a RM/0 operation. Three (*'s) are shown in Figure 12 since the RM/0 operation is now a slice operation. Figure 12A shows one middle edge cube positioned correctly. Figure 12B inverts the back side. Figure 12C rotates the middle layer one Q turn to the left. Figure 12D restores the right side (re-inverts it). Figure 12E corrects the middle layer by moving it one Q turn to the right. The RM/0 operation is done on either of the four faces (f, e, k, or r). Fig. 12 happens to be on the right 'r' face and assumes the 'f,r' edge cube is oriented correctly, it really only needs to be positioned correctly.

R/0//R/0//R/M

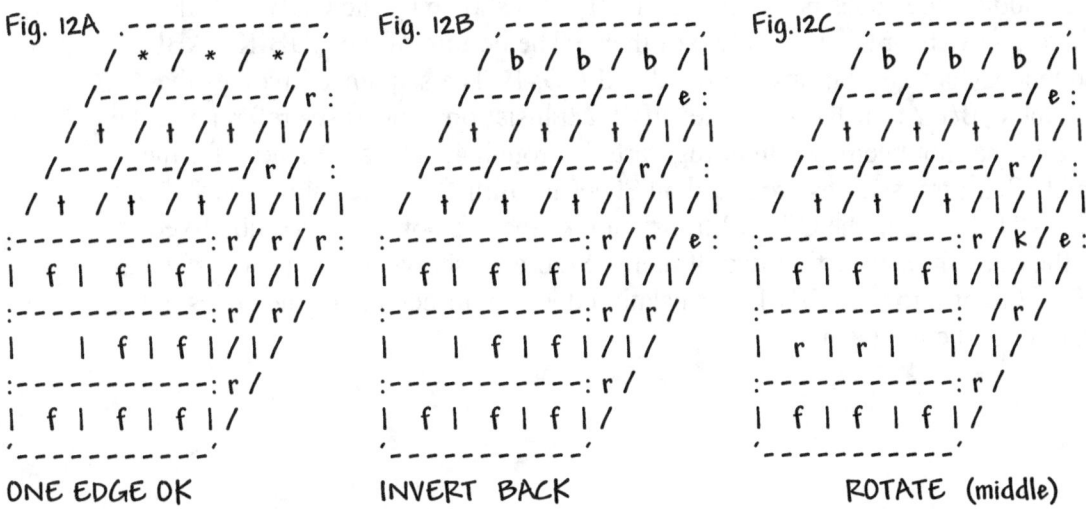

Fig. 12A   ONE EDGE OK   Fig. 12B   INVERT BACK   Fig. 12C   ROTATE (middle)

```
Fig. 12D .-----------,          Fig. 12E ,-----------,      If you have messed up the top or
         / * / * / * / |                 / * / * / * / |    bottom layers when trying to do
        /---/---/---/ r :               /---/---/---/ r :   RM/0, immediately go back and
       / † / † / † / 1 / 1            / † / † / † / 1 / 1    restore the top using the FOUR
      /---/---/---/ r /  :           /---/---/---/ r /  :    STEP APPROACH. Then restore
     / † / † / † / 1 / 1 / 1        / † / † / † / 1 / 1 / 1  the bottom as in Parts 1 and 2.
    :-----------: r / k / r:        :-----------: r / r / r: It is highly probable that you
    | f | f | f | f |/|/|/          | f | f | f | f |/|/|/   have only disturb the edge cubes
    :-----------: / r /            :-----------: r / r /     and need Part 2. If the bottom
    | r | r | r | 1 |/1/            |   | f | f | f |/1/      corners are wrong, back to Part 1.
    :-----------: r /              :-----------: r /         Remember, when mistakes happen,
    | f | f | f |/                  | f | f | f |/           going back to correct them gives one
    '-----------'                   '-----------'           plenty of practice with these algorithms.
    RESTORE BACK                    CORRECT (middle)
```

Part 4 Summary

Part 3 essentially moves the middle edge cubes in pairs and either positions one of them correctly or all four correctly (just positioned correctly, they don't have to be oriented correctly). Part 5 orients them. If only one is correct, then Part 4 uses one or two RM/0 slice operations to positioned all of them correctly (when one of the four middle edge cubes happened to be correct by chance or made correct by Part 3).

* * You are now about 40 twists away from making the CUBE whole. * *

SOLUTION Part 5
(Orients the last two or four middle edge cubes.)

Make sure your CUBE has its top layer cubes OK, all the bottom layer cubes OK, and the four middle edge cubes positioned correctly. As shown by the CUBIK TABLE, ALGO#1 inverts (twists or re-orients) two of them. The bottom of the CUBIK TABLE lists the sequence of six translations used to do ALGO#1. The sequence involves the R/V/1/1 and the C/B/1/2 translations. The C/B/1/2 translation which uses reference cube 3, see Figure 11, has not been used in doing Parts 1 through 4. The seven steps for the R/V/1/1 and C/B/1/2 translations are listed on Sheet 11 of the CUBIK subroutines. As shown by the CUBIK table, the C/B/1/2 translation keeps one bottom edge cube fixed and rotates the other three inverting two of them. Figure 13 shows the first three steps of doing a C/B/1/2 translation. A C/B/1/2 translation does not affect any of the cubes in the middle layer nor in the top layer.

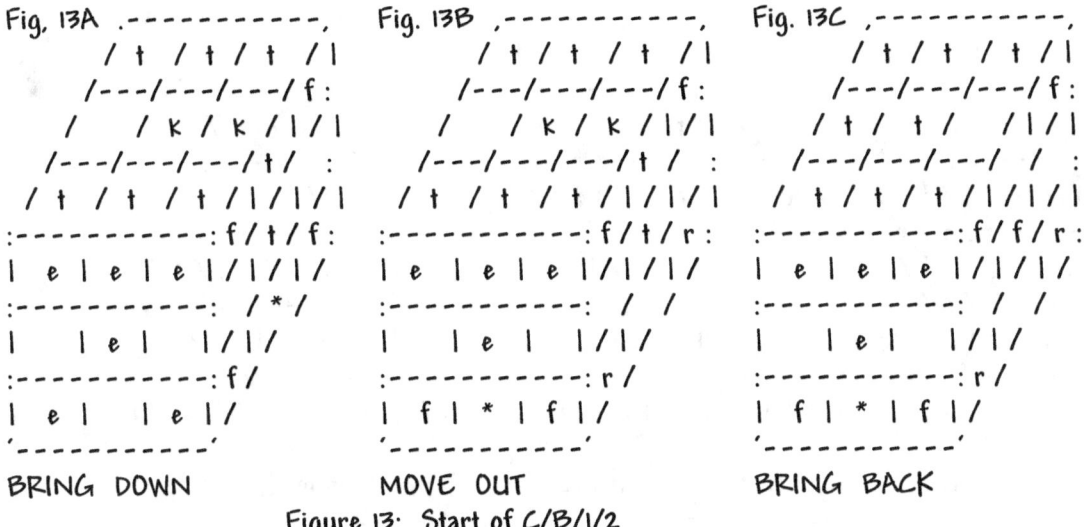

```
Fig. 13A .----------,      Fig. 13B .-----------,      Fig. 13C .-----------,
    / † / † / † /|            / † / † / † /|            / † / † / † /|
   /---/---/---/ f:          /---/---/---/ f:          /---/---/---/ f:
  /   / k / k /|/|          /   / k / k /|/|          / † / † /   /|/|
 /---/---/---/ † /  :      /---/---/---/ † /  :      /---/---/---/   /  :
/ † / † / † /|/|/|        / † / † / † /|/|/|        / † / † / † /|/|/|
:----------: f/†/f:       :----------: f/†/r:       :----------: f/f/r:
| e| e | e |/|/|/         | e | e| e |/|/|/         | e| e | e |/|/|/
:----------: /*/          :----------: / /          :----------: / /
|   |e |  |/|/            |   |e |  |/|/            |   |e |  |/|/
:----------: f/           :----------: r/           :----------: r/
| e|   | e |/             | f| * | f |/             | f| * | f |/
'----------'              '----------'              '----------'
BRING DOWN                 MOVE OUT                   BRING BACK
```
Figure 13: Start of C/B/1/2

Figurer 13A shows bringing down the center slice one Q turn. Figure 13B shows the bottom layer being rotated one Q turn to the left. Figure 13C shows bringing back the center slice. Step 4, not shown, rotates the bottom layer two Q turns to the left. Step 5 again brings down the center slice. Step 6 rotates the bottom layer one more Q turn so it looks like Figure 13A. Step 7 recovers the top by bringing back the center slice. (This operation is somewhat similar to the operation shown in Figure 10 except there the middle layer is rotated instead.)

Before starting ALGO#1 the two edge cubes to be inverted are to be on the edges facing you as shown in Figure 14 and for this example happen to be the 'f,r' and the 'f,e' cubes. If they were diagonally opposite instead, rotating one side of the CUBE two Q turns will place them both on the same face as shown in Figure 15 (here the 'f,r' and the 'e,k' cubes need to be inverted, so edge cubes 'f,e' and 'k,r' must be OK). Once in a blue moon, all four middle edge cubes need to be inverted (disoriented, but in their correct positions), then the ALGO#1 operation needs to be done twice, a worst case scenario). What ALGO#1 does, as indicated by Sheet 12 of the CUBIK subroutine flowchart, is to send these two cubes to the bottom layer using the R/V/1/1 translation, flip them over using the C/B/1/2 translation, and then bump them back to the middle layer using the R/V/1/1 translation which ends the CUBIK solution. The (*) cube in Figures 14, 15, and 16 is a reminder to use reference cube 2 when doing the R/V/1/1 translation. First do a R/V/1/1 translation on the front face which sends the 'f,r' cube to the bottom layer and the CUBE will look like Figure 16. Notice that the 'f,e' cube is now on the right edge of the right face, so next doing another R/V/1/1 translation with the right face towards you sends the 'f.e' cube to the bottom layer. Now hold the CUBE so the 'f,r' and 'f,e' cubes now on the bottom layer are at the left rear. With the CUBE in this position (left face should be towards you) do a C/B/1/2 translation which moves and flips the 'f,r' and 'f,e' cubes in the bottom layer. Now all that remains is to do three R/V/1/1 translations which correctly positions and orients three bottom edge cubes and the CUBE is solved. If your CUBE was like Figure 15, the 'f,r' and the 'e,k' cubes would be on the bottom instead.

```
Fig. 14    .-----------,     Fig. 15    ,-----------,     Fig. 16    ,-----------,
          / t / t / t /|               / b / t / t /|               / t / * / t /|
         /---/---/---/ r :            /---/---/---/ r :            /---/---/---/ r :
        / t / t / * /|/|             / b / t / * /|/|             / t / t / t /|/|=#
       /---/---/---/ r / r :        /---/---/---/ r / r :        /---/---/---/ r / f :
      / t / t / t /|/|/|           / b / t / t /|/|/|           / t / t / t /|/|/|
     :-----------: r / r / r :    :-----------: r / r / r :    :-----------: r / r / r :
     | f | f | f |/|/|/|         | k | f | f |/|/|/|         | f | f | f |/|/|/|
     :-----------: f / r /       :-----------: f / r /       :-----------: k / r /=f,r
     | e | f | r |/|/|      k,e=| e | f | r |/|/|           | e | f | r |/|/|
     :-----------: r /           :-----------: r /           :-----------: r /
     | f | f | f |/              | k | f | f |/              | f | f | f |/   #= f,e
     '-----------'               '-----------'               '-----------'

   (f,e) & (f,r) not oriented   (k,e) & (f,r) not oriented   (f,r) on bottom; (f,e) is next
```

ALGO#1 PREPARATION

As indicated by the subroutine, the first thing to do is to rotate the middle layer minus one Q turn (rotate one Q turn to the left) which prepares the CUBE to send down the first of three edge cubes to the bottom layer. Now look at the four middle edge cubes very carefully, there exists one and only one edge cube in the middle layer that is in a READY position to be transferred correctly to the bottom layer using a R/V/1/1 translation (refer back to Figure 6A if necessary). After locating this cube, do a R/V/1/1 translation with this cube on the right front edge. One cube is done, two to go. *(If you are generating a complicated pattern you need to know that the second edge cube that was sent to the bottom layer is now the one being bump back.)* To prepare for the next cube, rotate the middle layer plus two Q turns (rotate two Q turns to the right). Again there is now one and only one middle edge cube that is in a READY position to be transferred correctly to the bottom layer using a R/V/1/1 translation. After locating this cube, do a R/V/1/1 translation with this cube on the right front edge. Two cubes are done, one to go (three of the four bottom edge cubes should be correct). *(If you are generating a complicated pattern you need to know that the bottom layer cube that is out of position, directly opposite to the first edge cube that was sent down, is now the one being bump back.)* No preparation is necessary for the third cube (it is already in a READY position), therefore, position the CUBE so the last cube to be transferred to the bottom layer is on the right front edge, then do a R/V/1/1 translation. *(If you are generating a complicated pattern you need to know that the first edge cube that was sent to the bottom layer is now the one being bump back.)* The bottom layer should now be complete and the four middle edge cubes fall out in their correct places making the CUBE one solid configuration again after correcting the middle layer minus one Q turn. If your CUBE was like Figure 15, two corrections are necessary in order to make the CUBE whole again. Correct the middle layer minus one Q turn and then re-invert the side.

```
END OF           .--------------------.
SOLUTION     / ttttt   / ttttt   / ttttt  / |
             / ttttt   / ttttt   / ttttt  / r |
             / ttttt   / ttttt   / ttttt  / rr |
          /------/------/------/ rrr  :
          / ttttt  / ttttt  / ttttt  / | rr / |
          / ttttt  / ttttt  / ttttt  / r | r / r |
          / ttttt  / ttttt  / ttttt  / rr | / rr |
       /------/------/------/ rrr / rrr  :
        / ttttt  / ttttt  / ttttt  / | rr / | rr / |
       / ttttt  / ttttt  / ttttt  / r | r / r | r / r |
     / ttttt  / ttttt  / ttttt  / rr | / rr | / rr |
   :--------------------: rrr / rrr /| rrr :
   |  fffff  |  fffff  |  fffff  | rr / | rr / | rr /
   |  fffff  |  fffff  |  fffff  | r / r| r / r | r /
   |  fffff  |  fffff  |  fffff  | / rr | / rr | /
   :--------------------: rrr / rrr /
   |  fffff  |  fffff  |  fffff  | rr / | rr /
   |  fffff  |  fffff  |  fffff  | r / r | r /
   |  fffff  |  fffff  |  fffff  | / rr | /
   :--------------------: rrr /
   |  fffff  |  fffff  |  fffff  | rr /
   |  fffff  |  fffff  |  fffff  | r /
   |  fffff  |  fffff  |  fffff  | /
   '--------------------'
```

Part 5 SUMMARY

I admit ALGO#1 is long, but it works. I admit that I felt real lucky that Saturday when after doing ALGO#1 for the first time the CUBE popped out whole. It's a game and I won. The CUBIK ALGORITHM TABLE and the CUBIK FLOWCHART have greatly helped me to explain the rules of the game.

If mistakes were made while doing ALGO#1, go back to Part 2, Part 3, or Part 4 if only edge cubes were messed up. Back to Part 1 if the bottom corners were inadvertently disturb, and if the top layer was disturb, back to the FOUR STEP APPROACH.

Here again I admit that my 1982 way of getting the bottom four corners 'in-sync' and the positioning of the four middle edge cubes was somewhat a haphazard adventure. The new 'out-of-sync' subroutine and the new RM/o slice operation has made these two operations precise. And it is to my chagrin that there no doubt is a reasonable slice operation out there somewhere to replace my ALGO#1.

(Dear Cubik World Continued)

Last March when I had three cubes, I often brought them to work configured with my favorite patterns. One of the engineers suggested that I make the pattern shown in Figure P31BB2'' of the pattern section. I made the pattern and as often the case I would show the pattern to Bill. Two days later he told me that he was able to do this pattern from a solid CUBE using the C/B/1/2 and C/B/-1/-2 translations that were two algorithms (translations) he used frequently. I call the pattern the 'Ring-around-the-corner' pattern and there are two versions of it as shown in Figures 18A and 18B. The strategy for doing them is discussed after explaining the C/B/-1/-2 translation that in effect is doing the C/B/1/2 translation in reverse.

29

```
   BEFORE        AFTER
:------------:  :------------:
|   | i |  |   |   | h |  |
:------------:  :------------:
| h |   | g |  | g |   | i |   C/B/1/2
:------------:  :------------:
|   | fix |  |   |   | fix |  |
:------------:  :------------:
                             (top
                            views)
:------------:  :------------:
|   | i |  |   |   | g |  |
:------------:  :------------:
| h |   | g |  | i |   | h |   C/B/-1/-2
:------------:  :------------:
|   | fix |  |   |   | fix |  |
:------------:  :------------:
```

The two diagrams to the left show the position of the four bottom edge cubes before and after the translations. The two edge cubes that get flipped during the translation are underlined.

Figure 17 shows the first three steps of doing a C/B/-1/-2 translation. The The seven steps are the same as for the C/B/1/2 translation listed on Sheet 11 of the Cubik Subroutine flowchart except you rotate the bottom layer to the right instead of to the left for the 2nd, 4th, and 6th steps.

Reference cube 3 is the asterisk cube in Figure 17A. STEP 1 brings down the center slice as in Figure 17A. The reference cube is moved out one Q turn to the right as in Figure 17B for STEP 2. Figure 17C shows the top restored minus the reference cube 3 for STEP 3. STEP 4, not shown, rotates the bottom layer two Q turns to the right. STEP 5 brings down the center slice again. STEP 6 rotates the bottom layer one Q turn so it looks like Figure 17A in order for STEP 7 to restore the top. Figure 17 represents 'steps' done on a solid CUBE.

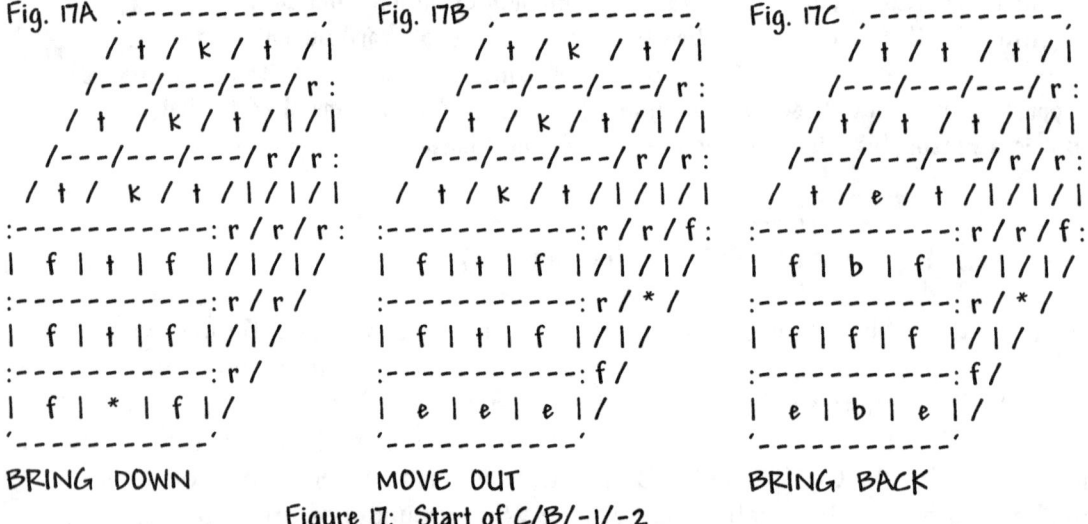

Figure 17: Start of C/B/-1/-2

This 'Ring-around-the corner' sequence only puts a ring around one corner where as the 'Ring-around-the-corner' pattern referred to in the pattern section which is made from a scrambled CUBE puts a 'Ring-around-the-corner' pattern on two diagonally opposite corners. However, by repeating the sequence, a 'Ring-around-the-corner' pattern can be put around two, three, or four corners. *(Just recently when I did the four corners from a scrambled CUBE I quickly realized that you can get lost in ALGO#1, therefore, I specifically identified the cube to be bumped back to the middle layer during the last three R/V/1/1 translations of ALGO#1).*

```
Fig. 18A .-----------,        Fig. 18B ,-----------,        Fig. 19  ,-----------,
        / b / b / b /|                 / b / b / b /|                 / r / b / r /|
       /---/---/---/ K :               /---/---/---/ K :               /---/---/---/ f :
      / b / b / e /|/|                 / b / b / K /|/|                 / r / r / r /|/|
     /---/---/---/ b / K:              /---/---/---/ e / K:              /---/---/---/ f / f:
    / b / e / b /|/|/|                 / b/ K / b /|/|/|                 / r/ r / r /|/|/|
   :-----------: K/ K/ K:             :-----------: K/ K/ K:             :-----------: f/ f/ f:
   | e | K | e |/|/|/               | e | b | e |/|/|/               | t | t | t |/|/|/
   :-----------: b/ K/              :-----------: e/ K/              :-----------: f/ f/
   | e | e | K |/|/                | e | e | b |/|/                | t | t | t |/|/
   :-----------: K/                :-----------: K/                :-----------: f/
   | e | e | e |/                  | e | e | e |/                  | t | t | t |/
   '-----------'                   '-----------'                   '-----------'
```

'Ring-around-the-corner' Patterns

Making a 'Ring-around-the-corner' pattern from a solid CUBE entails doing two C/B/1/2 translations and one C/B/-1/-2 translation. Doing these translations observing the top, right, and front faces of the CUBE results in the pattern appearing around the bottom left back corner. To do this sequence you first remember the top, right, and front colors and then do a C/B/1/2 translation with that front color towards you. Secondly, orient the CUBE so the right color is now on top, the top color is towards you, and the front color is now on the right, as shown in Figure 19, and then do another C/B/1/2 translation. Without reorienting the position of the CUBE, rotate the back layer of the CUBE two Q turns and then do a C/B/-1/-2 translation (with the top color still towards you) which results in the pattern shown in Figure 18A after you correct the back slice two Q turns. If you repeat this sequence again, starting with the top color on the top, right color on the right, and front color towards you as you did previously, you will end up with the other phase of this pattern as shown in Figure 18B. Hence, if you repeat the sequence again, the CUBE will return to a solid configuration again. This 'Ring-around-the-corner' pattern happens to be cyclic in three. If you repeat the sequence starting with different colors as the reference colors, the pattern will appear on a different corner, however, check before starting that the corner patterns will not interfere with each other.

At this point in my 'cubik' story, I must make a confession. Last night I purchased another cubic book. I looked through it twice during the Holidays and couldn't resist buying it any longer. It's the 82-pattern book by Taylor and Rylands. It's their wondrous algorithms for making all those patterns from a solid CUBE that amazes me so, frankly, I never thought it was possible. I soon tried their algorithm for the 'Bit Mason' arrangement and from a solid CUBE ended up with the pattern of Figure P30A in the pattern section of this article then without changing the position of the CUBE I repeated the algorithm and ended up with the pattern of Figure P30B, and after doing it once more it resolved into a solid CUBE again, consequently, it is a cyclic algorithm having a cycle of three, a total of 36 twists of the CUBE. I guess you have to conclude that all the legitimate algorithms in their book are cyclic in two or in three cycles, I'll be mentioning this again in the pattern section. With regards to patterns from a solid CUBE, I never spent much effort doing this, however, I would always pick up a scrambled CUBE lying around the house and often made it into a solid pattern and left it that way. One evening I saw it had turned into a DOT pattern and I asked my daughter how did she do it and she retorted, 'Dad, I kept making a lot of H's', well I stayed up a little extra that night learning how to make a DOT pattern from a solid CUBE which is explained in the pattern section.

About a month after solving the CUBE I traced the cubic diagram off the front cover of the Scientific American magazine in order to make several copies (blanks). I then located my daughter's old box of crayons and ended up making some 25 colored pictures as I generated different pattern configurations. It was relaxing to crayon again and it was a way to keep a record of which group of patterns I had generated. I also used my daughter's new VG-1 camera to take pictures of my three CUBES and the crayon pictures. I eventually plan to display all these pictures in my train room. Before Easter I volunteered to play a minor role in a play at the church for Palm Sunday. It was a fun part. The play was called 'A Parade Without a Permit' and the staging look like a business office. The night before the dress rehearsal I asked the director if I could place my three 'Rubik' CUBES around the stage and she replied that it would be okay. Well the three CUBES just looked beautiful up there on the stage with the five spotlights bringing out their brilliant colors. One CUBE was on the edge of a desk, another on a bookcase, and the third at the edge of a table. When the play was over, the cast had to deliver palm leafs to the audience immediately after the curtain call. As soon as I made my round with the leaves I went back to the stage and there they were, three very young boys each with a 'Rubik's CUBE twisting away. I just stayed off to the side and watched them. Eventually their parents located them and the boy's reluctantly parted with their CUBE. I could tell by the expression on the parent's faces that they had no idea where their children went. I didn't say a word and heard the parents say that they thought the CUBEs belong to the church. This episode is additional proof why Mr. Rubik's invention is such a success. It appeals to the young as a toy, but to the rest a game, a puzzle, a menace, or a Hungarian horror.

In June my youngest daughter had her sixteenth surprise birthday party. I still had my three CUBEs and left them sitting on the piano in the living room. There were some two-dozen girls there and I was busy helping my wife earlier in the evening. After nine o'clock the girls started picking up the CUBEs and some did succeed in solving one side. One and half hours later I went into the room where they were with the CUBEs and sat down amongst them and started to give them hints as to how to solve the CUBE. They were very fascinated with the CUBEs and I had a lot of fun watching them churn away until after midnight. They caught onto the idea of the 7 or 8 step sequences very fast and with a little guidance enjoyed correctly placing the cubes one by one.

Later in the summer our family attended a party given by one of the ladies in my wife's study group. About twelve families get together and it gives our daughters a chance to meet the other young people since they are all home from college. I still had two CUBEs (finally gave the other one to my brother) and took them along and left them with my youngest daughter. I was with several men in the TV room, one was the host, a surgeon, a couple of lawyers, etc. and as usual there is always a bottle of wine on the table and the conversation revolved around wine tasting. For the past year most of them had been attending a wine tasting school and I enjoyed listening to them. All the wines taste the same to me, anyway. Soon our host's twenty-year old daughter came into the room with one of my CUBEs and said to me as she was handing me the CUBE 'Do it'. Well, I handed it back to her and said that I would let her do it. She consented and sat down on the floor beside me. I felt confident that without touching the CUBE I could instruct her in how to solve it since I had a similar experience at my daughter's sixteenth birthday party. My doctor and lawyer friends always enjoy having an engineer in their midst and really enjoyed watching my engineering approach to solving the CUBE as she started to unscramble the CUBE. Once in a while I would ask her to show me one of the sides in order to check whether she moved the reference cube the prescribed number of Q turns (either one or two). Other than that, she did make the comment after a while that solving the CUBE was getting to be a little boring since there are many repetitions of the 7 or 8 step sequences, but they work. After the CUBE turned out solid again, they applauded my logical approach in being able to guide someone through unscrambling the Rubik's CUBE which really made me feel great. A party at my brother's home, about twenty relatives and neighbors, gave me another chance to be the center of attention when several nieces asked me to unscramble the CUBE for them. They were between the ages of six and ten and enjoyed twisting the solid CUBE into different configurations until it became too mixed up then they would ask me to make it whole again.

Other times during the year when I enjoyed being a CUBEMEISTER was when going to parades on a bus. I enjoyed showing fellow band members the different patterns that could be configured. It was a good way to make the time go fast while going to and from the parades. My last episode with the CUBE was when visiting the minister that married us. I saw that he has a couple of CUBEs lying on the living room table and inquired about them and found out that he had them for sometime and enjoyed picking them up and dabbles with them a little bit not getting too far towards unscrambling them. I told him about my adventure in writing this article and indicated that after church that evening I would talk him through solving the CUBE. I found out that he didn't have the tiny instruction booklet that comes with the Ideal CUBE package and he didn't realize the significance of the center cubes being used as reference cubes since they were fixed to a spindle. As inferred in the beginning of this article, the CUBE is really a big enigma for anyone who doesn't read the tiny instruction book that comes with the CUBE or never gets any advice about the mechanical nature of the CUBE. The minister enjoyed twisting his way towards seeing the CUBE gradually becoming unscrambled and remarked at the end that 'It works'. I later attempted to disassemble one of his CUBEs. They were both stiff since they were not used that much. Well as I was trying to pry the first cube loose I heard a crack. Well, from now on when I want to show anyone a disassembled CUBE, I'll make sure that it's mine. At least the replacement CUBE sent to him will have the tiny instruction booklet. This concludes my Cubik story for one enjoyable Cubik year. My biggest thrill has been to put a CUBE in people's own hands when they ask me to unscramble it for them and talk them through the Cubik solution.

I made the Cubik flowcharts by using a vertical bar that is part of the type with the Diablo printer. I also used the overlay feature that is part of the word processor. In generating this article it was easy with the word processor to move CUBE diagrams or flowcharts symbols from place to place.

The pattern section describes how to make a DOT pattern from a solid CUBE with four twists (moves). Also, an IMAGE DOT pattern is shown made from a solid CUBE with four twists (moves). With six twists (moves), a DOT/BAR pattern is made. These three types of patterns are in a way cyclic since it is possible to go from a solid, to a pattern, another pattern, still another pattern, etc., and back to a solid pattern again. In generating this story and making all these patterns I was always a little apprehensive about discovering some simpler ending algorithms to my Cubik solution, but none surfaced. It still takes me a good seven minutes to solve the CUBE using this algorithm approach. I really enjoy making patterns from a scrambled CUBE using the Cubik solution.

I would like to thank my manager and supervisor for allowing me to put the CUBE on the TUBE, my fellow engineers who thought it was a good idea, my daughter for reviewing this 'cubik' cookbook solution, and my wife for reviewing my 'cubik' grammar.

Last year at work I had a 500-page report to do (half tables and half text) and my manager has two secretaries type it using the word processor. Late in December I started to review it on the tube and got familiar with the word processor. After New Years on my lunch hour I experimented designing CUBE pictures with the word processor that led to trying to describe my solution that led to an article and finally this short story. I guess that I have spent some 100-hours on the tube with this story plus another 100-hours with the flowcharts and patterns section. After work and on Saturdays I spent a couple of hours entering the text and would make corrections on my lunch hour. We have a high-speed printer (only prints caps) and at home I would review what I had done using a copy from this printer. The final copy is from a DIABLO printer.

After writing this story I feel that I could instruct someone to solve the CUBE over the telephone (a TV hookup would be helpful), of course they would have to chip in on the phone bill.

CUBE-fully yours,
Honabe.

March, 1982

P.S.   Attached is a CUBIK ALGORITHM Table, a CUBIK FLOWCHART, some comments on CONFIGURATIONS (PATTERNS), and a FOUR STEP APPROACH review. Last, I better thank Lance for instituting our word processing and helping me to overcome my 'cubik' hang-ups at the keyboard, and Clancy for the flowchart graphics. Also, to Rich who was always there. *Added to the PATTERN section is the cycling of the seven algorithms plus three exercises (examples).*

BUT, I got carried away with the idea of identifying the flowchart paths being used and added seven more examples. NOW, being more efficient I quickly added ten more examples in an addendum then finally one "post-post" example.

FLOWCHART

INTRODUCTION

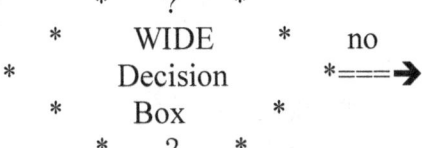

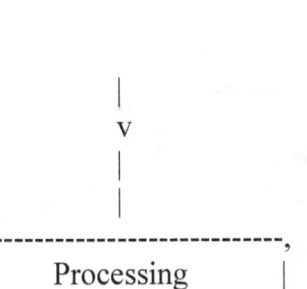

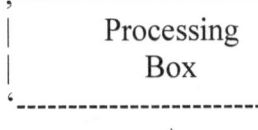

There are perhaps many methods to present a solution to the Rubik's CUBE. Flowcharting is a graphical means of representation the logical steps of the solution and it is effective in providing an overview of the logical flow of a solution which enables further evaluation of alternative approaches. Hopefully, since the problem was solved before the flowchart was generated, there are no flaws or "bugs" in the routine.

Once completing the CUBIK ALGORITHM TABLE, the task of flowcharting the CUBIK solution was both challenging and enjoyable.

A flowchart is basically a collection of boxes and lines. The boxes indicate what is to be done and the lines indicate the sequence of the boxes. The boxes are of various shapes which represent the action to be performed in the routine.

The rectangle processing box is an instruction to carry out a given task. Sometimes a brief description of the task to be performed is included within the symbol.

A diamond is used to indicate a point in the routine where a choice is made on the flow of the routine from the point. A test condition is included within the symbol and the possible results of this test are used to label the respective flows from the symbol. In this routine they are always "yes" or "no". The WIDE decision box is capable of having more words.

The double-line box is used to indicate a point in a main routine that calls a sub-routine. Upon reaching a sub-routine box leave the main routine and follow the sub-routine flowchart. The end of a sub-routine has the word "RETURN" which means to return to the main routine and continue on from that point.

Sub-routines are beneficial when a certain number of logical steps are used several times in the solution. Putting these often used steps in a sub-routine means only flowcharting them once. This reduces the size of the flowchart and in most cases simplifies the routine (or solution). A sub-routine can call out a sub-sub-routine, etc. A sub-sub-routine must return to the sub-routine prior to returning to the main routine.

CUBE = whole 'cube'.
cube = pieces of the whole 'cube' (cubelets).
OK = cublet(s) which are both positioned and oriented correctly.

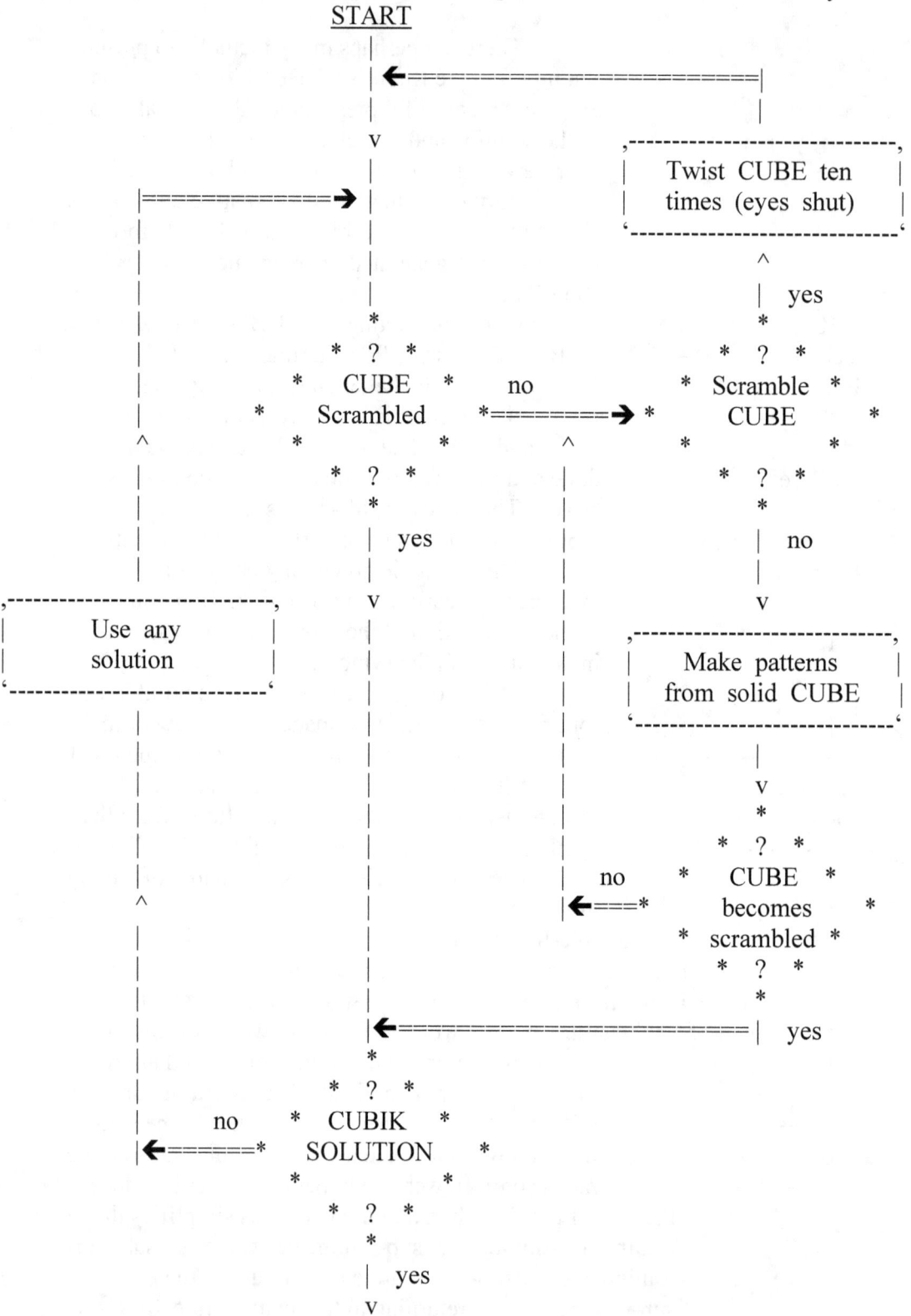

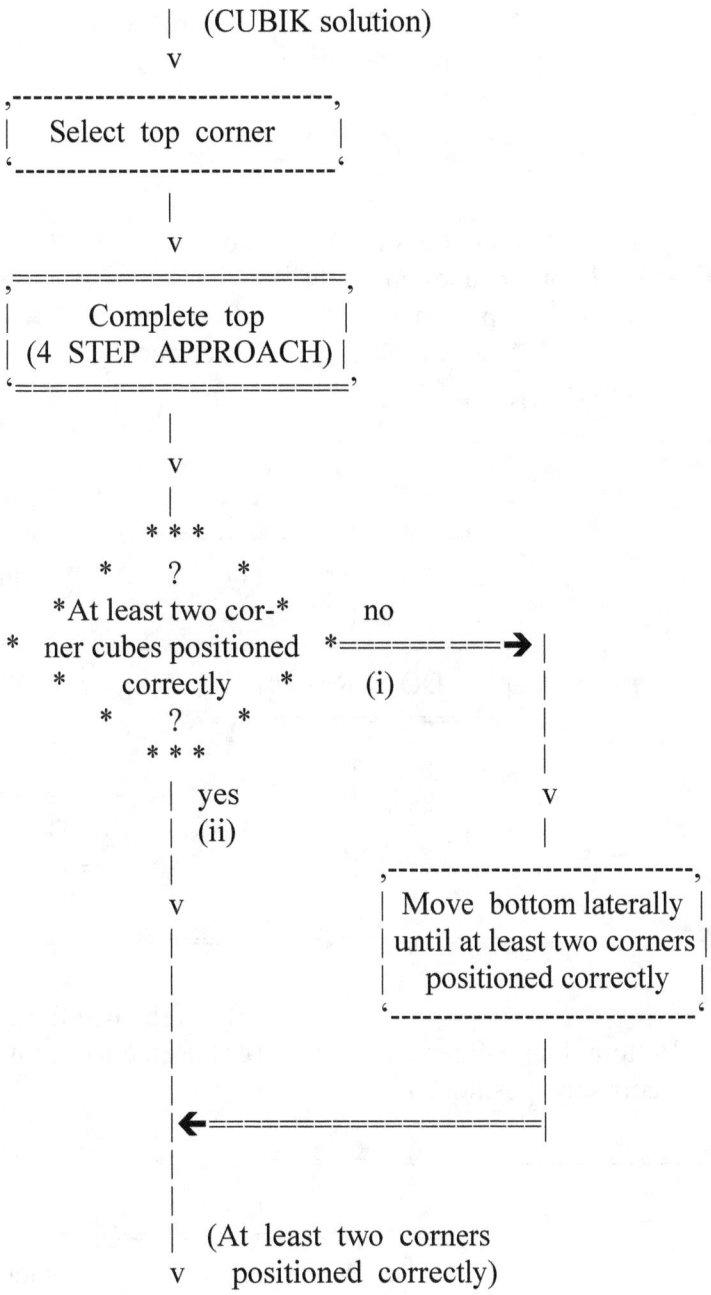

```
                    |  (CUBIK solution)
                    v
        ,---------------------------,
        |   Select  top  corner     |
        '---------------------------'
                    |
                    v
        ,===================,
        |    Complete  top  |
        | (4 STEP APPROACH) |
        '==================='
                    |
                    v
                    |
                * * *
              *    ?    *
            *At least two cor-*          no
          *  ner cubes positioned *=====  ===➔|
            *    correctly    *      (i)       |
              *    ?    *                       |
                * * *                           |
                  | yes                          |
                  | (ii)                         v
                  |                              |
                  v                              |
                  |              ,---------------------------,
                  |              | Move  bottom laterally |
                  |              | until at least two corners |
                  |              |  positioned correctly   |
                  |              '---------------------------'
                  |                              |
                  |                              |
                  |⬅==================|
                  |
                  |
                  |  (At  least  two  corners
                  v     positioned  correctly)
```

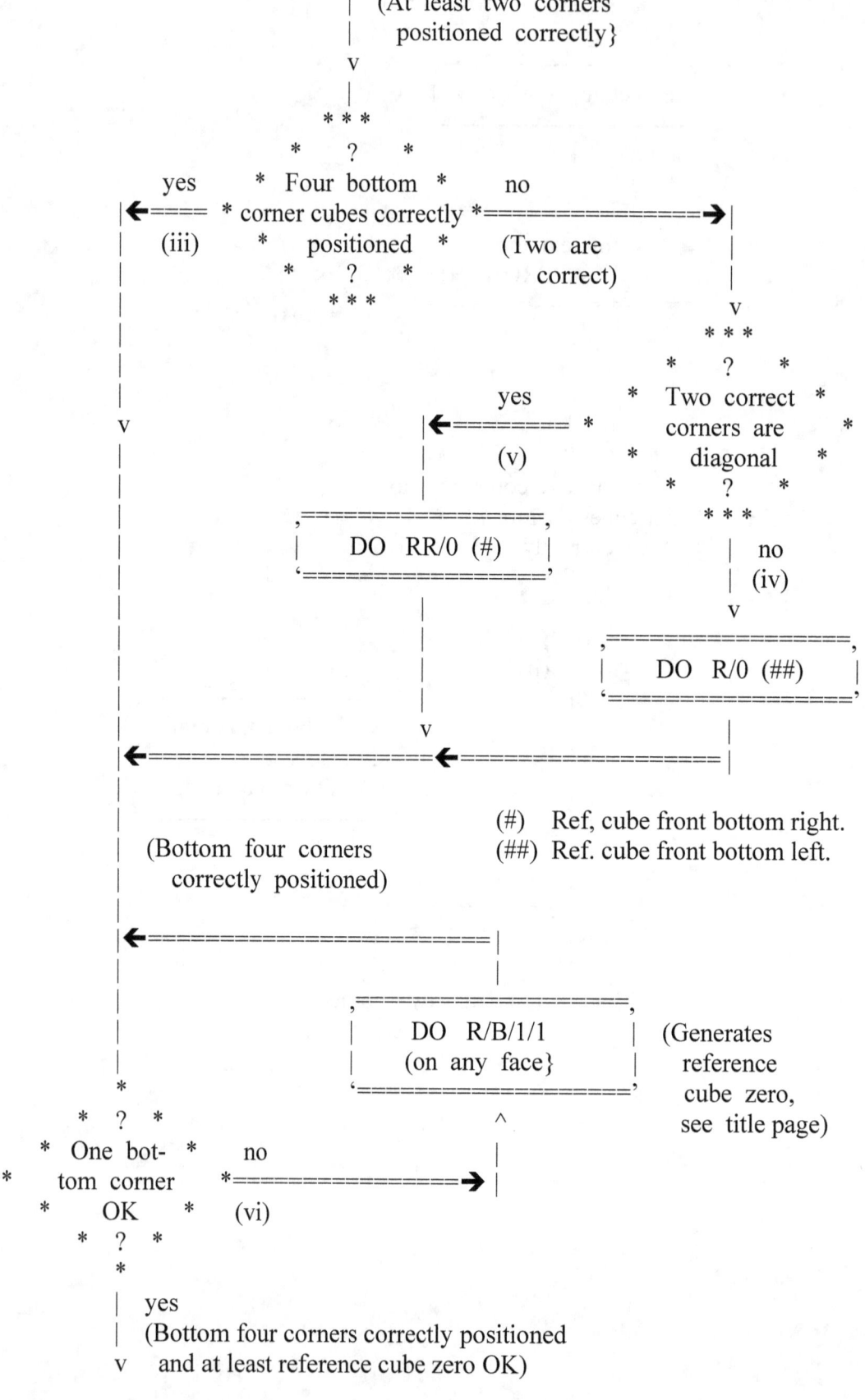

(#)   Ref, cube front bottom right.
(##) Ref. cube front bottom left.

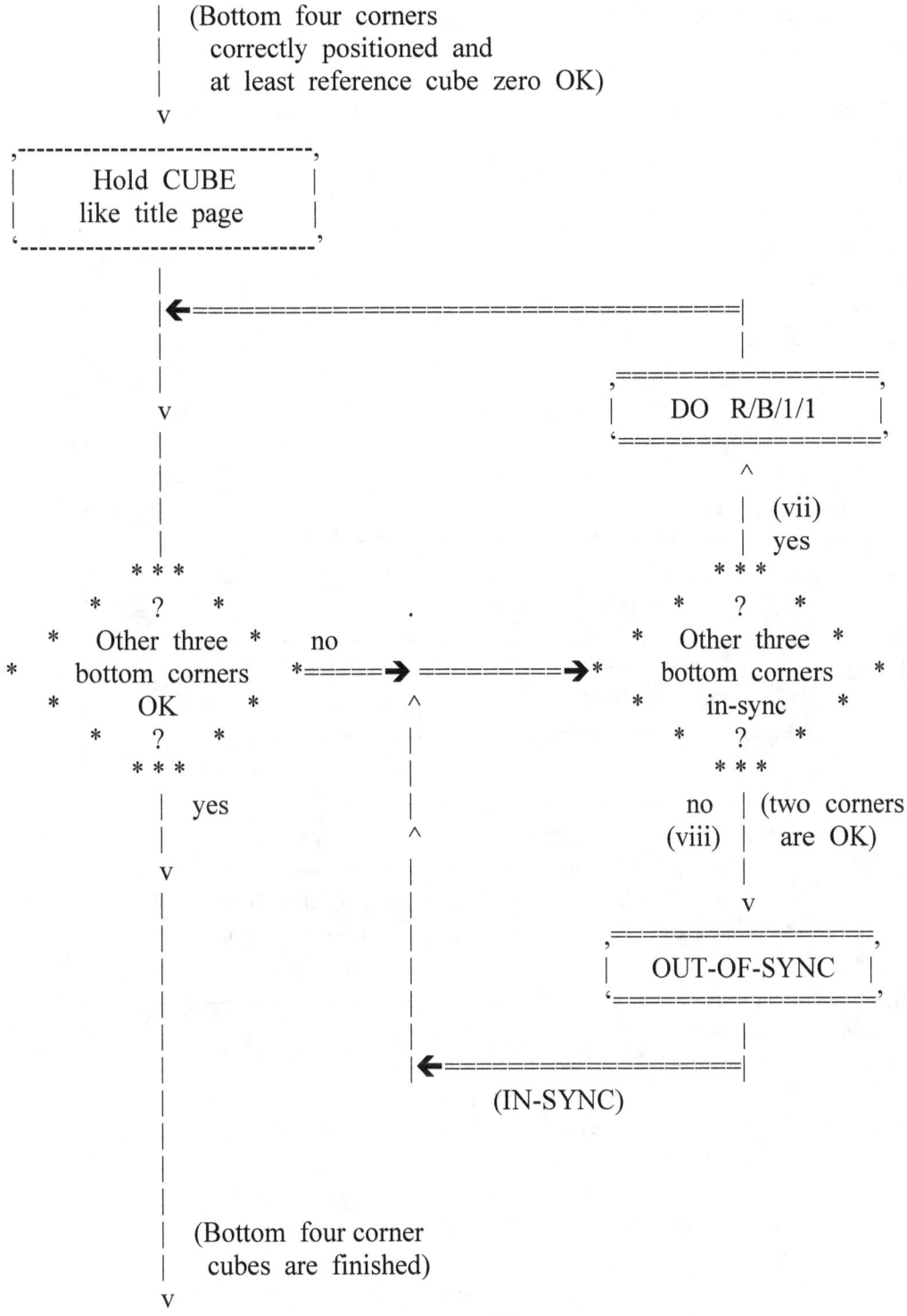

| (Bottom four corners
| correctly positioned and
| at least reference cube zero OK)
v
,------------------------------,
| Hold CUBE |
| like title page |
'------------------------------'
        |
        |◄==|
        | |
        | ,==================,
        v | DO R/B/1/1 |
        | '=================='
        | ∧
        | | (vii)
        | | yes
      * * *   * * *
      *   ?   *   *   ?   *
   * Other three *   no   * Other three *
* bottom corners *====➤ ==========➤ * bottom corners *
   *   OK   *   ∧   *   in-sync   *
      *   ?   * | *   ?   *
      * * * | * * *
        | yes | no | (two corners
        | ∧   (viii) | are OK)
        v | |
        | | v
        | | ,==================,
        | | | OUT-OF-SYNC |
        | | '=================='
        | | |
        | |◄================|
        | (IN-SYNC)
        |
        |
        | (Bottom four corner
        | cubes are finished)
        v

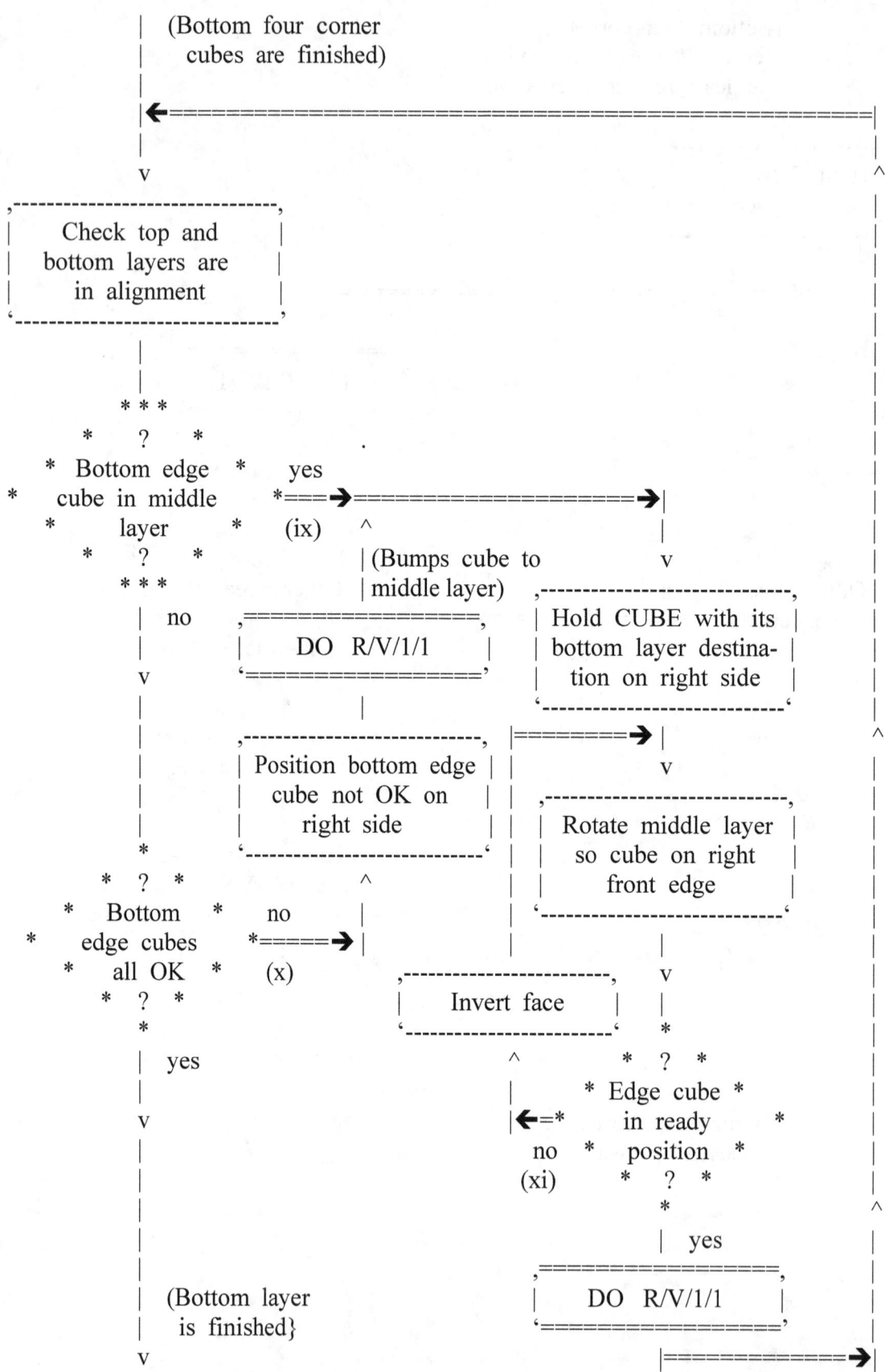

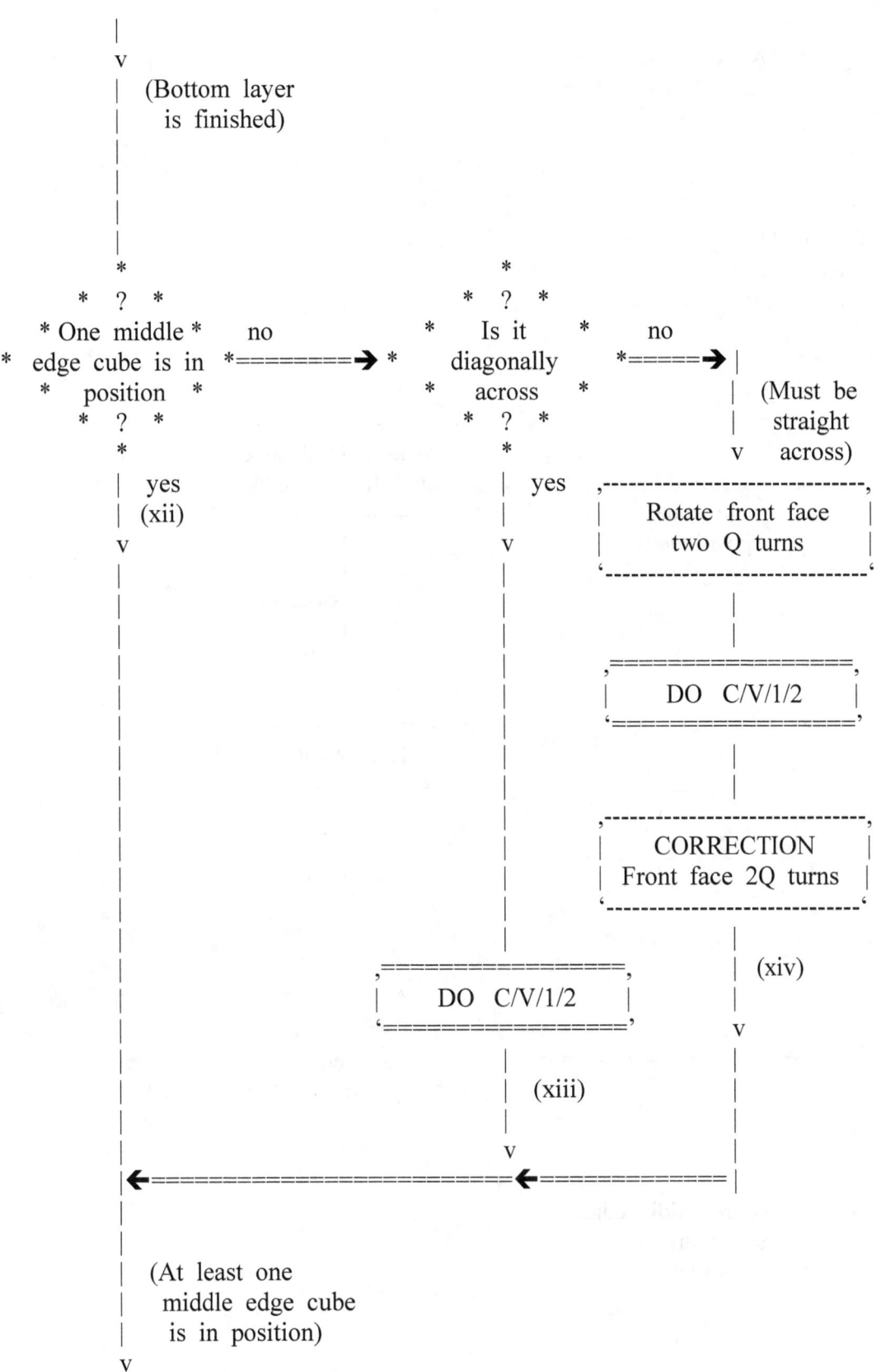

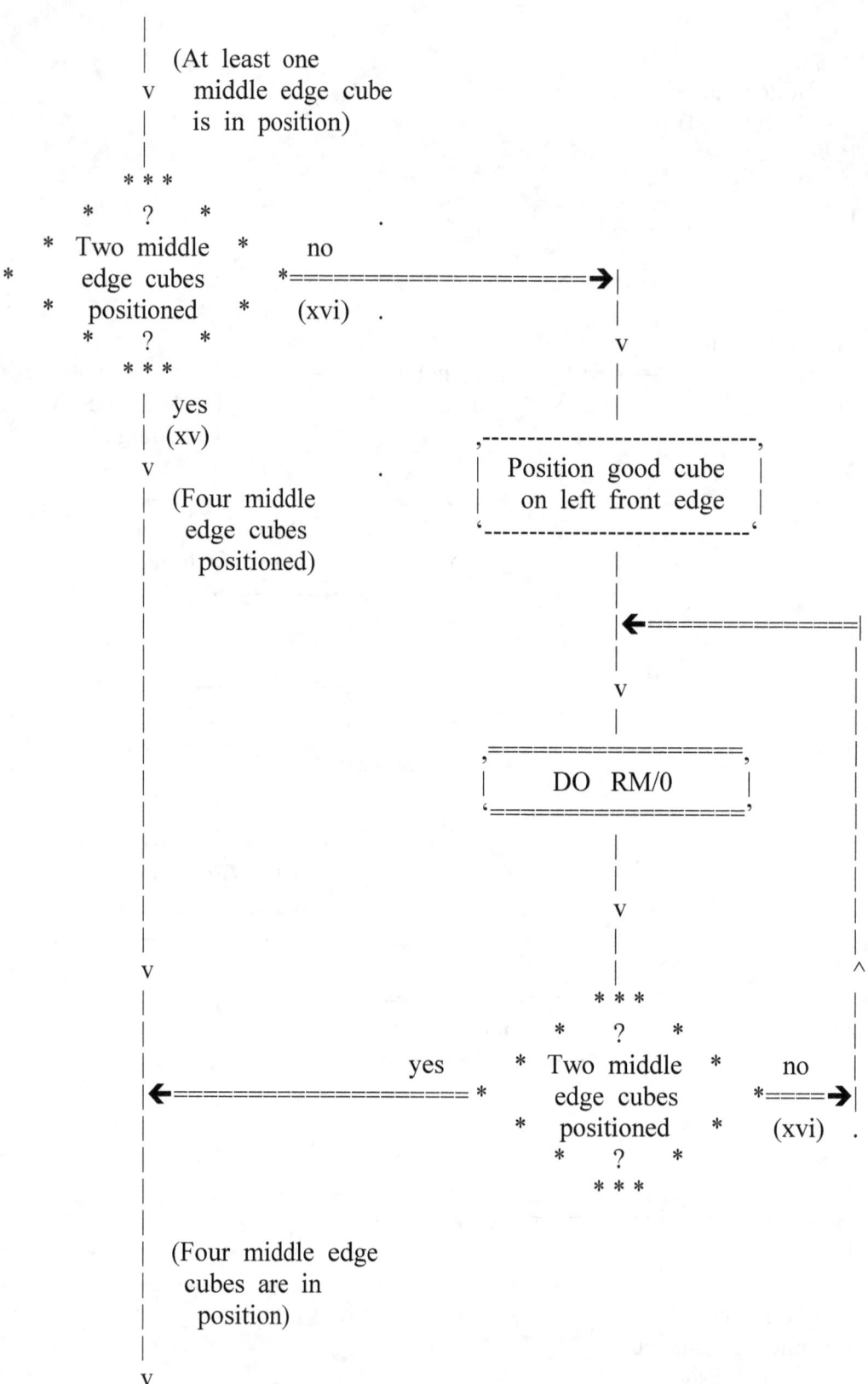

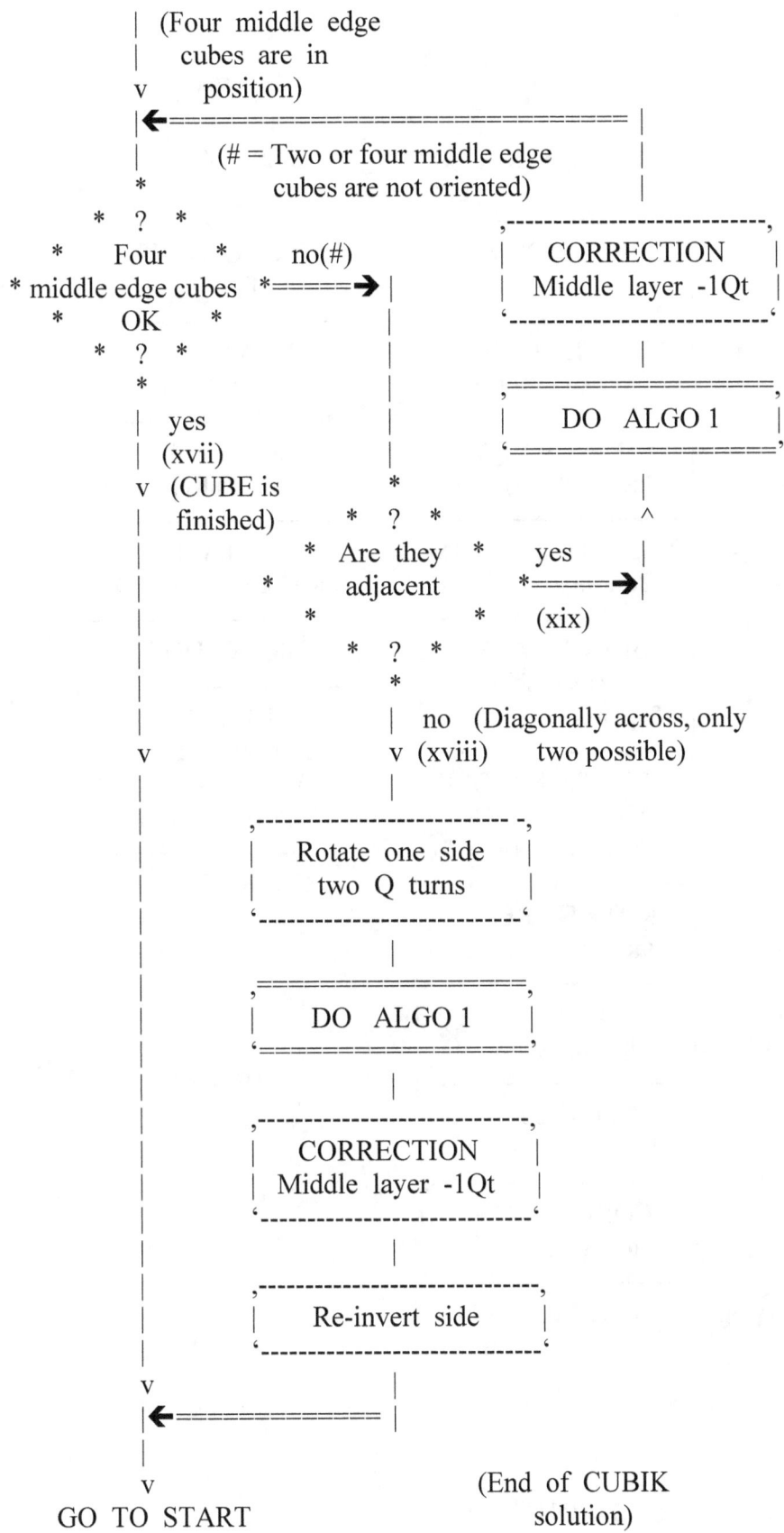

```
      | (Four middle edge
      |   cubes are in
      v    position)
      |←===============================|
      |         (# = Two or four middle edge    |
      *             cubes are not oriented)      |
   *  ?  *                                       |
  *  Four  *      no(#)      ,--------------------,
 * middle edge cubes *=====→ |      CORRECTION     |
  *   OK   *                 |  Middle layer -1Qt  |
   *  ?  *                   '--------------------'
      *                      |                  |
      |  yes                 |        ,==================,
      |  (xvii)              |        |    DO  ALGO 1    |
      v  (CUBE is     *      |        '=================='
      |   finished)   *  ?  *|                  |
      |             *  Are they *   yes          ^
      |              *  adjacent *====→|         |
      |               *         *   (xix)        |
      |                *  ?  *                    |
      |                   *                       |
      |                   | no  (Diagonally across, only
      v                   v (xviii)    two possible)
      |                   |
      |          ,-------------------,
      |          |  Rotate  one  side  |
      |          |    two  Q  turns    |
      |          '-------------------'
      |                   |
      |          ,==================,
      |          |    DO   ALGO 1   |
      |          '=================='
      |                   |
      |          ,-------------------,
      |          |     CORRECTION     |
      |          |  Middle  layer -1Qt |
      |          '-------------------'
      |                   |
      |          ,-------------------,
      |          |    Re-invert side   |
      |          '-------------------'
      v                   |
      |←=============|
      |
      v                (End of CUBIK
 GO  TO  START            solution)
```

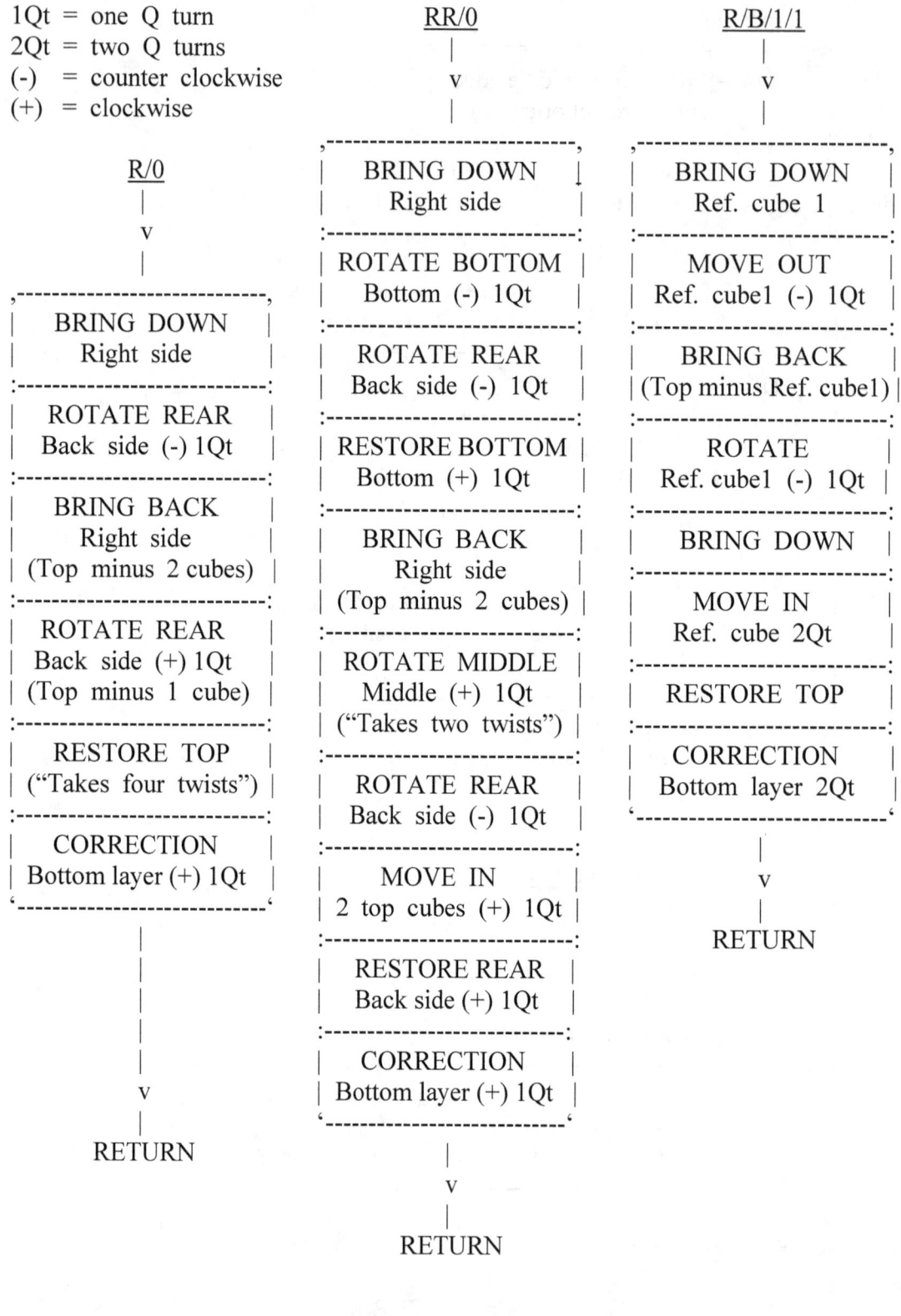

1Qt = one Q turn
2Qt = two Q turns
(-) = counter clockwise
(+) = clockwise

R/0
|
v
|

```
,---------------------------,
|      BRING DOWN           |
|      Right side           |
:---------------------------:
|     ROTATE REAR           |
|   Back side (-) 1Qt       |
:---------------------------:
|     BRING BACK            |
|     Right side            |
|  (Top minus 2 cubes)      |
:---------------------------:
|     ROTATE REAR           |
|   Back side (+) 1Qt       |
|   (Top minus 1 cube)      |
:---------------------------:
|     RESTORE TOP           |
|  ("Takes four twists")    |
:---------------------------:
|     CORRECTION            |
|  Bottom layer (+) 1Qt     |
'---------------------------'
```
|
|
|
|
|
v
|
RETURN

RR/0
|
v
|

```
,---------------------------,
|      BRING DOWN           |
|      Right side           |
:---------------------------:
|    ROTATE BOTTOM          |
|    Bottom (-) 1Qt         |
:---------------------------:
|     ROTATE REAR           |
|   Back side (-) 1Qt       |
:---------------------------:
|   RESTORE BOTTOM          |
|    Bottom (+) 1Qt         |
:---------------------------:
|     BRING BACK            |
|     Right side            |
|  (Top minus 2 cubes)      |
:---------------------------:
|    ROTATE MIDDLE          |
|    Middle (+) 1Qt         |
|  ("Takes two twists")     |
:---------------------------:
|     ROTATE REAR           |
|   Back side (-) 1Qt       |
:---------------------------:
|       MOVE IN             |
|  2 top cubes (+) 1Qt      |
:---------------------------:
|    RESTORE REAR           |
|   Back side (+) 1Qt       |
:---------------------------:
|     CORRECTION            |
|  Bottom layer (+) 1Qt     |
'---------------------------'
```
|
v
|
RETURN

R/B/1/1
|
v
|

```
,---------------------------,
|      BRING DOWN           |
|      Ref. cube 1          |
:---------------------------:
|      MOVE OUT             |
|  Ref. cube1 (-) 1Qt       |
:---------------------------:
|     BRING BACK            |
| (Top minus Ref. cube1)    |
:---------------------------:
|      ROTATE               |
|  Ref. cube1 (-) 1Qt       |
:---------------------------:
|     BRING DOWN            |
:---------------------------:
|      MOVE IN              |
|   Ref. cube 2Qt           |
:---------------------------:
|    RESTORE TOP            |
:---------------------------:
|     CORRECTION            |
|  Bottom layer 2Qt         |
'---------------------------'
```
|
v
|
RETURN

[SUB_ROUTINES]

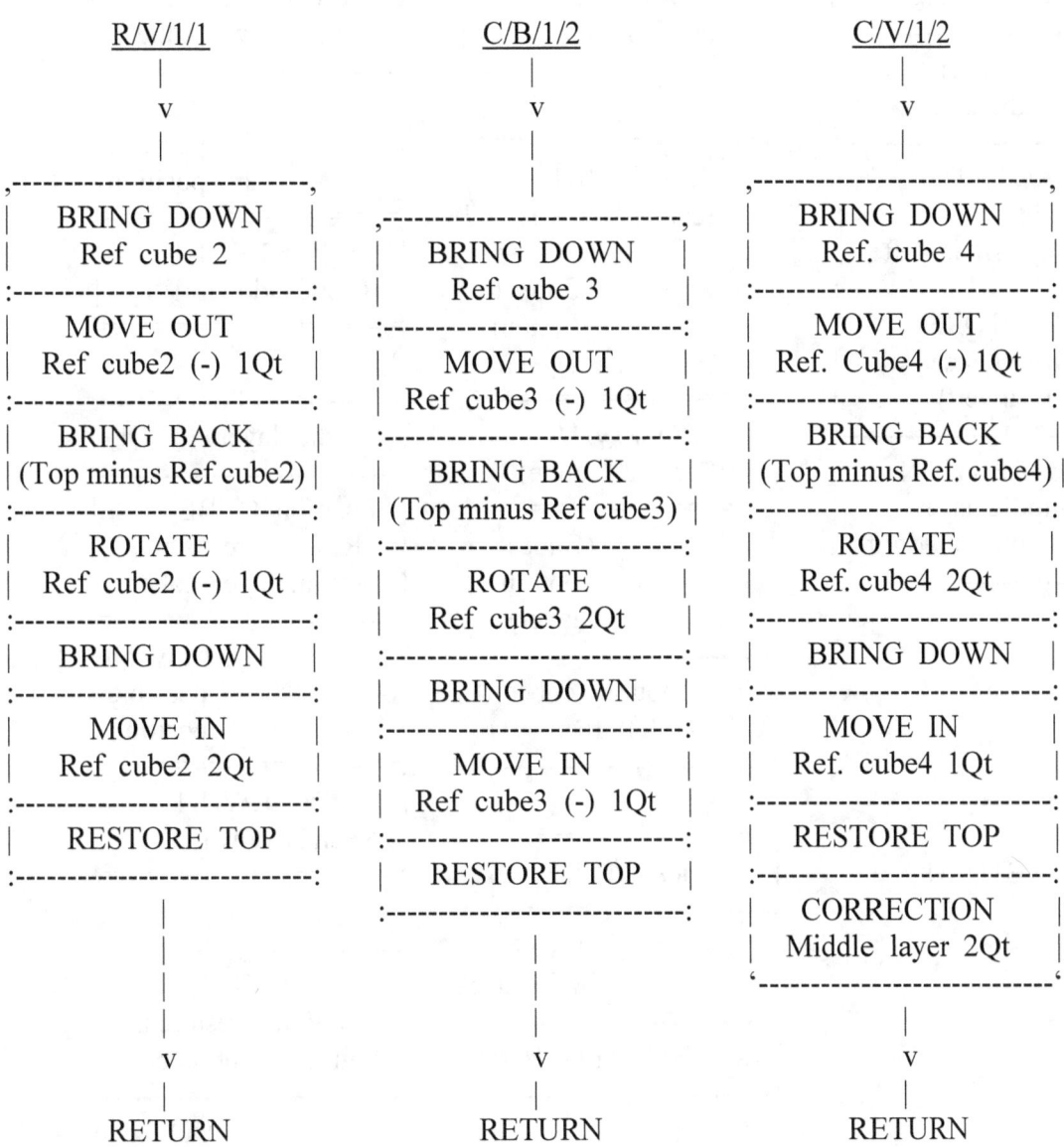

R/V/1/1

v

```
,----------------------,
| BRING DOWN           |
| Ref cube 2           |
:----------------------:
| MOVE OUT             |
| Ref cube2 (-) 1Qt    |
:----------------------:
| BRING BACK           |
| (Top minus Ref cube2)|
:----------------------:
| ROTATE               |
| Ref cube2 (-) 1Qt    |
:----------------------:
| BRING DOWN           |
:----------------------:
| MOVE IN              |
| Ref cube2 2Qt        |
:----------------------:
| RESTORE TOP          |
:----------------------:
```

v

RETURN

C/B/1/2

v

```
,----------------------------,
| BRING DOWN                 |
| Ref cube 3                 |
:----------------------------:
| MOVE OUT                   |
| Ref cube3 (-) 1Qt          |
:----------------------------:
| BRING BACK                 |
| (Top minus Ref cube3)      |
:----------------------------:
| ROTATE                     |
| Ref cube3 2Qt              |
:----------------------------:
| BRING DOWN                 |
:----------------------------:
| MOVE IN                    |
| Ref cube3 (-) 1Qt          |
:----------------------------:
| RESTORE TOP                |
:----------------------------:
```

v

RETURN

C/V/1/2

v

```
,----------------------------,
| BRING DOWN                 |
| Ref. cube 4                |
:----------------------------:
| MOVE OUT                   |
| Ref. Cube4 (-) 1Qt         |
:----------------------------:
| BRING BACK                 |
| (Top minus Ref. cube4)     |
:----------------------------:
| ROTATE                     |
| Ref. cube4 2Qt             |
:----------------------------:
| BRING DOWN                 |
:----------------------------:
| MOVE IN                    |
| Ref. cube4 1Qt             |
:----------------------------:
| RESTORE TOP                |
:----------------------------:
| CORRECTION                 |
| Middle layer 2Qt           |
`----------------------------´
```

v

RETURN

[SUB_ROUTINES]

Flowchart

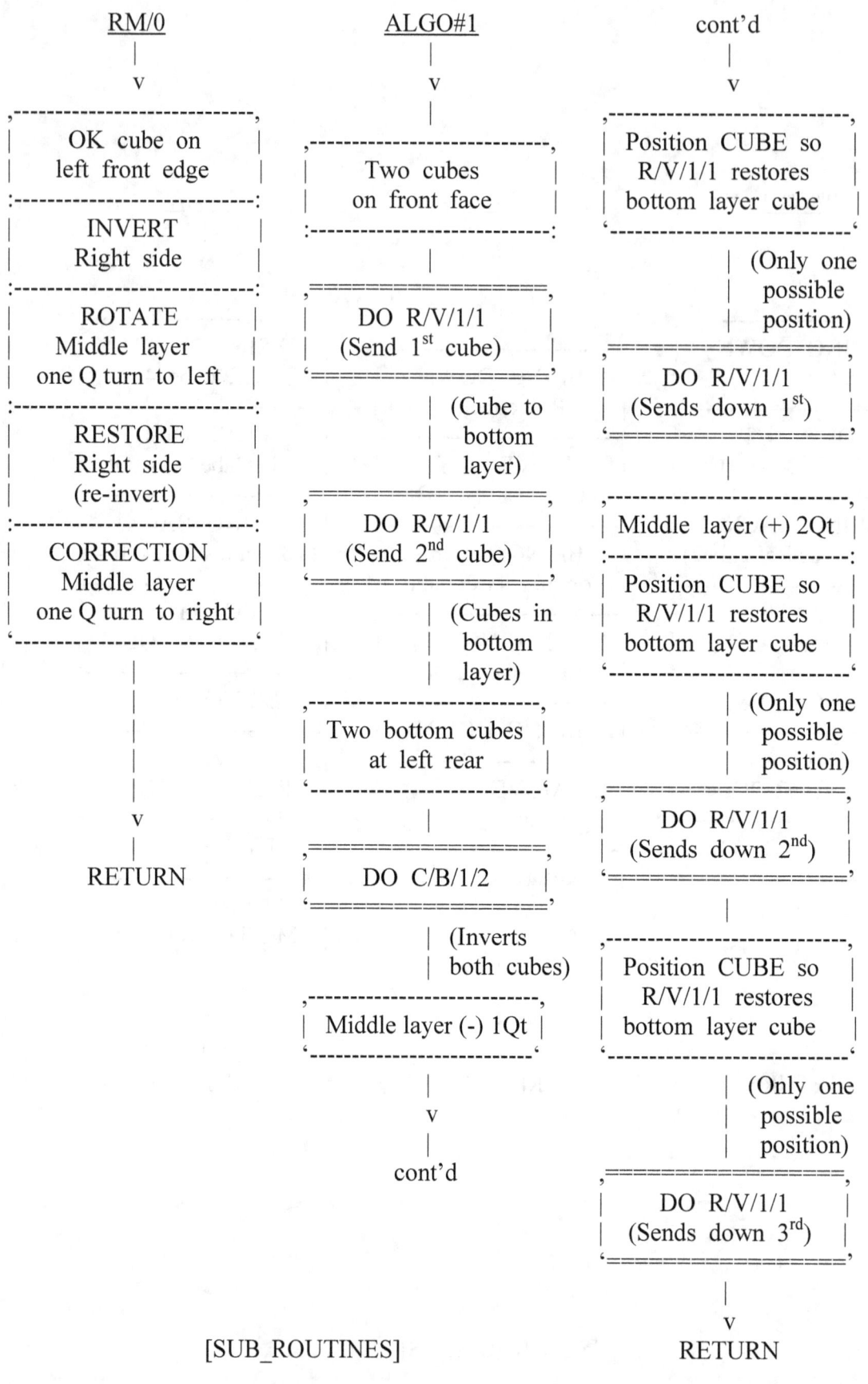

RM/0

↓
v

,----------------------------,
| OK cube on |
| left front edge |
:----------------------------:
| INVERT |
| Right side |
:----------------------------:
| ROTATE |
| Middle layer |
| one Q turn to left |
:----------------------------:
| RESTORE |
| Right side |
| (re-invert) |
:----------------------------:
| CORRECTION |
| Middle layer |
| one Q turn to right |
'----------------------------'

|
|
|
|
|
v
|
RETURN

ALGO#1

↓
v
|

,----------------------------,
| Two cubes |
| on front face |
:----------------------------:
|
================,
| DO R/V/1/1 |
| (Send 1st cube) |
'================'
| (Cube to
| bottom
| layer)
,================,
| DO R/V/1/1 |
| (Send 2nd cube) |
'================'
| (Cubes in
| bottom
| layer)
,----------------------------,
| Two bottom cubes |
| at left rear |
'----------------------------'
|
,================,
| DO C/B/1/2 |
'================'
| (Inverts
| both cubes)
,----------------------------,
| Middle layer (-) 1Qt |
'----------------------------'
|
v
|
cont'd

cont'd

↓
v

,----------------------------,
| Position CUBE so |
| R/V/1/1 restores |
| bottom layer cube |
'----------------------------'
| (Only one
| possible
| position)
,================,
| DO R/V/1/1 |
| (Sends down 1st) |
'================'
|
,----------------------------,
| Middle layer (+) 2Qt |
:----------------------------:
| Position CUBE so |
| R/V/1/1 restores |
| bottom layer cube |
'----------------------------'
| (Only one
| possible
| position)
,================,
| DO R/V/1/1 |
| (Sends down 2nd) |
'================'
|
,----------------------------,
| Position CUBE so |
| R/V/1/1 restores |
| bottom layer cube |
'----------------------------'
| (Only one
| possible
| position)
,================,
| DO R/V/1/1 |
| (Sends down 3rd) |
'================'
|
v
RETURN

[SUB_ROUTINES]

Flowchart   Sheet 12 of 13

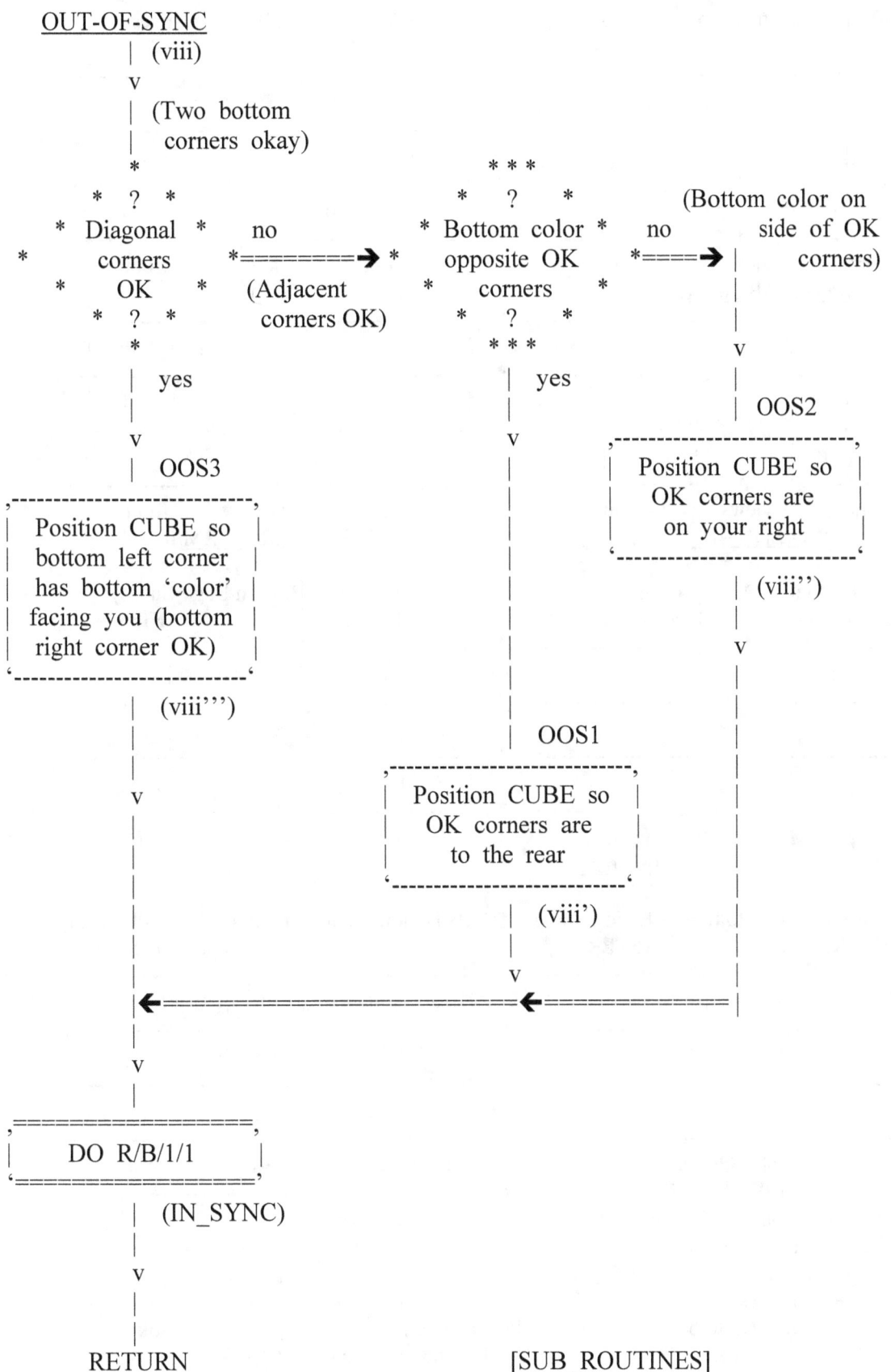

OUT-OF-SYNC

RETURN

[SUB_ROUTINES]

(Positions Bottom Four Corners)
  BEFORE   AFTER

```
,------------------,      ,------------------,
| fix |     | y  |       | fix |     | x  |
:------------------:      :------------------:
|     |     |     |       |     |     |     |
:------------------:      :------------------:
| x   |     | FIX |       | y   |     | FIX |
'------------------'      '------------------'
        front      top        front
                  views
```

R/0//B/0//Bk/0 = RR/0

```
,------------------,      ,------------------,
| v   |     | w  |       | w  |     | v  |
:------------------:      :------------------:
|     |     |     |       |     |     |     |
:------------------:      :------------------:
| FIX |     | fix |       | FIX |     | fix |
'------------------'      '------------------'
        front      top        front
                  views
```

R/0//Bk/0 = R/0

*   *   *   *   *   *   *

(Orients Bottom   *   (ENDING
Four Corners)   Inverts j and k)

```
,------------------,      ,------------------,
| tws |     | tws |       |     |     |     |
:------------------:      :------------------:
|     |     |     |       | j   |     | k   |
:------------------:      :------------------:
| FIX |     | tws |       |     |     |     |
'------------------'      '------------------'
       front                  front  face
```

R/B/1/1   *   ALGO #1

*   *   *   *   *   *

(Positions/Orients Bottom Four Edge Cubes)
  BEFORE   AFTER

```
   ,------------------,     ,------------------,
 f |   |   |   |   |    f |   |   |   |   |
 r :------------------:   r :------------------:
 o | u |   |   |   |    o |   |   |   |   |
 n :------------------:   n :------------------:
 t |   |   |   |   |    t |   | u |   |   |
   '------------------'     '------------------'
```

views of right face
R/V/1/1

FIX = stays positioned and oriented
Q = ¼ turn   tws = twist
(?) = cube inverted   (ALGO #1)
  1) R/V/1/1, (k) to bottom
  2) R/V/1/1, (j) to bottom
  3) C/B/1/2, inverts j & k
  4) Middle minus 1Qt

(Positions Middle Edge Cubes)
  BEFORE   AFTER

```
,------------------,      ,------------------,
| m |     | n |         | p |     | o |
:------------------:      :------------------:
|     |     |     |       |     |     |     |
:------------------:      :------------------:
| o   |     | p   |       | n   |     | m   |
'------------------'      '------------------'
        front      top        front
                  views
```

C/V/1/2

```
,------------------,      ,------------------,
| t |     | s |         | r |     | t |
:------------------:      :------------------:
|     |     |     |       |     |     |     |
:------------------:      :------------------:
| FIX |     | r   |       | FIX |     | s   |
'------------------'      '------------------'
        front      top        front
                  views
```

R/0//R/0//M/0 = RM/0

*   *   *   *   *   *   *

(Shifts Three Bottom Edge Cubes)
  BEFORE   AFTER

```
,------------------,      ,------------------,
|   | i |   |   |       |   | (h) |   |
:------------------:      :------------------:
| h |     | g |         | g |     | (i) |
:------------------:      :------------------:
|   | FIX |   |         |   | FIX |   |
'------------------'      '------------------'
        front      top        front
                  views
```

C/B/1/2

*   *   *   *   *   *

*(Gets Bottom Four Corners In-Sync @R/B/1/1)
  OOS1   b OOS2

```
,------------------,      ,------------------,
| ok |     | ok |       |     |     | ok |
:------------------:      :------------------:
|     |     |     |       |     |     |     |
:------------------:      :------------------:
|     |     |     |       |     |     | ok |
'------------------'      '------------------'
        front      top        front
     b   front  b       views  b   front
```

(When two corners
  are OK)

```
           ,------------------,
           | ok |   |   |<b
           :------------------:
           |   |   |   |   | O
           :------------------: O
           |   |   | ok |   | S
           '------------------' 3
                 b   front
```

5) R/V/1/1, send down 1st
6) Middle plus 2Qt's
7) R/V/1/1, send down 2nd
8) R/V/1/1, send down 3rd

Cubik Algorithms

CUBIK ALGORITHM TABLE REVIEW

a}   Doing a RR/0 operation on the front face of the CUBE towards you causes two diagonal opposite bottom corner cubes to be interchanged. The 'FIX' bottom right front corner cube remains positioned and oriented correctly (is OK) but the 'fix' bottom left rear corner cube gets re-oriented.

b}   Doing a R/0 operation on the front face of the CUBE towards you causes two adjacent bottom corner cubes to be interchanged. The 'FIX' bottom left front corner cube remains positioned and oriented correctly (is OK) but the 'fix' bottom right front corner cube gets re-oriented. (NOTE: three consecutive R/0 operations can replace the RR/0 operation in (a) above as shown below,)

<u>3 R/0's = RR/0 (top views)</u>

BEFORE	AFTER R/0(1)	ADJUST +1Qt
fix | | y	y | | fix	x | | y
| | |	| | |	| | |
x | | FIX	x | | FIX	FIX | | fix

AFTER R/0(2)	ADJUST (-)Qt	AFTER R/0(3)
y | | x	x | | fix	fix | | x
| | |	| | |	| | |
FIX | | fix	y | | FIX	y | | FIX

c}   Doing an R/B/1/1 operation on the front face of the CUBE towards you causes the bottom left front corner cube to remain 'FIXED' and the other three bottom corner cubes get re-oriented, twisted. When the other three bottom corner cubes are IN-SYNC, one or two R/B/1/1 operations cause all four of the bottom corner cubes to be positioned and oriented correctly (are OK).

d}   When two of the four bottom corner cubes are positioned and oriented correctly (are OK), OOS1, OOS2, and OOS3 puts three of the bottom corner cubes IN-SYNC by doing an R/B/1/1 operation on the front face so that a task 'c' operation above causes all four bottom corner cubes to be OK. 'b' represents the color of the bottom face. OOS1, OOS2, and OOS3 are the three possible arrangements of the bottom four corner cubes when two of the four corner cubes are OK, when none are OK doing an R/B/1/1 operation arbitrarily on any one of the four side faces results in eight possible outcomes, four are IN-SYNC outcomes and four have two corners that are OK (see page 17).

e}   Doing an R/V/1/1 operation on the front face of the CUBE towards you causes the right front middle edge cube to be sent to the bottom right edge cube position where it is positioned and oriented correctly (is OK). An R/V/1/1 operation only churns the four middle edge cubes. The top face and the bottom four corner cubes are not disturbed.

Algorithm Review

After several R/V/1/1 operations the top face and the bottom face are now completed, all that remains is to position and orient the four middle edge cubes.

f}   Doing a C/V/1/2 operation on any face of the CUBE towards you causes the four middle edge cubes to be interchanged diagonally. This operation is used to position correctly at least one, two, or perhaps four middle edge cubes. This operation does not disturb either the top face or the bottom face.

g}   When one of the three middle edge cubes is positioned correctly (doesn't have to be correctly oriented), doing a RM/0 operation on the front face of the CUBE towards you causes three of the four middle edge cubes to be rotated. One or two RM/0 operations cause all four middle edge cubes to be correctly positioned. This operation does not disturb either the top face or the bottom face.

h}   Doing an ALGO #1 operation starting with the front face of the CUBE towards you inverts two of the middle edge cubes in order to orient them. Sometimes all four middle edge cubes are OK and don't need to be oriented, then again all four might need to be oriented then two ALGO #1 operations are required. *WOW!, at last a CUBE is in its pristine state.*

i}   In the midst of an ALGO #1 operation, doing an C/B/1/2 operation on the front face of the CUBE towards you causes three of the four bottom edge cubes to be rotated but the two edge cubes at the left and rear get inverted during the translation. This is the operation that inverts the 'ending' two (or sometimes four) middle edge cubes after they are sent to the bottom layer by two separate R/V/1/1 operations. A C/B/1/2 operation only disturbs edge cubes in the bottom layer. After the C/B/1/2 routine in the ALGO #1 operation which disturbs three edge cubes in the bottom layer, it naturally then requires three separate R/V/1/1 operations to restore the three bottom disturbed edge cubes but they are restored in such an orderly fashion that the middle edge cubes are also restored at the same time completing the CUBIK solution.

j}   As a reminder the diagrams below show when the bottom corner cubes are IN-SYNC as a result of task 'd' above.

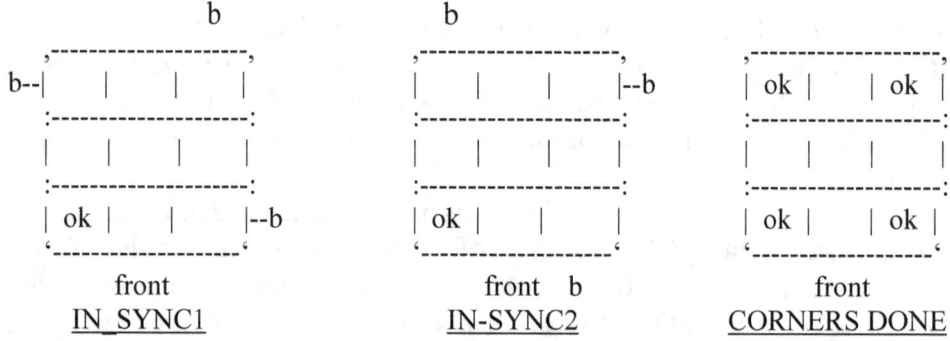

b		b		

IN_SYNC1   IN-SYNC2   CORNERS DONE

When the bottom corners are IN-SYNC they will appear either as IN-SYNC1 or as IN-SYNC2. Doing a R/B/1/1 operation on the IN-SYNC1 configuration results in the IN-SYNC2 configuration. Doing a R/B/1/1 operation on the IN-SYNC2 configuration results in the bottom CORNERS DONE configuration. 'b' represents the color of the bottom face.

FOUR STEP APPROACH (flowchart)
    CUBE = whole 'cube'
    cube = pieces of the whole CUBE (cublets)
    Qt = quarter turn

COMPLETE TOP

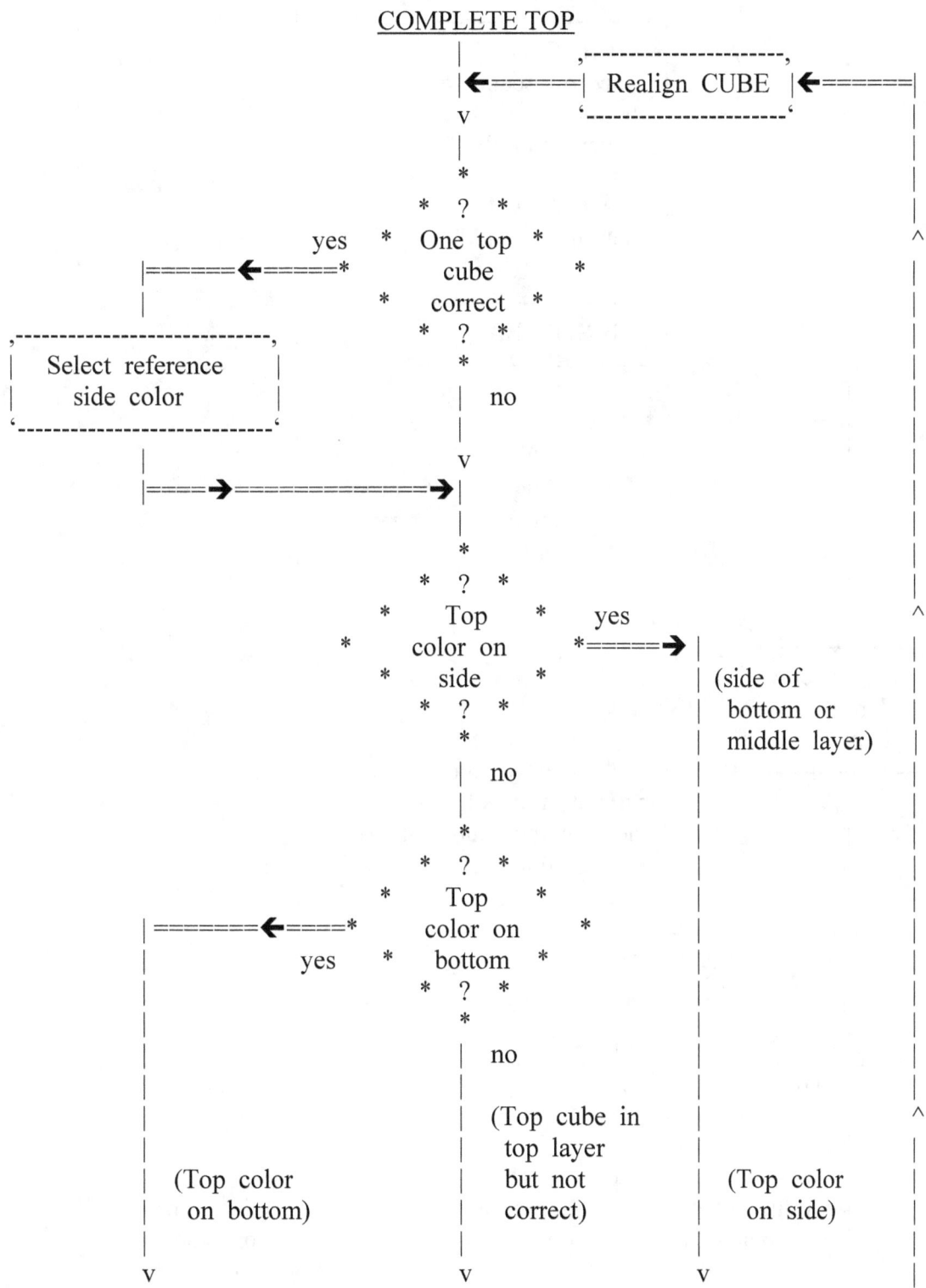

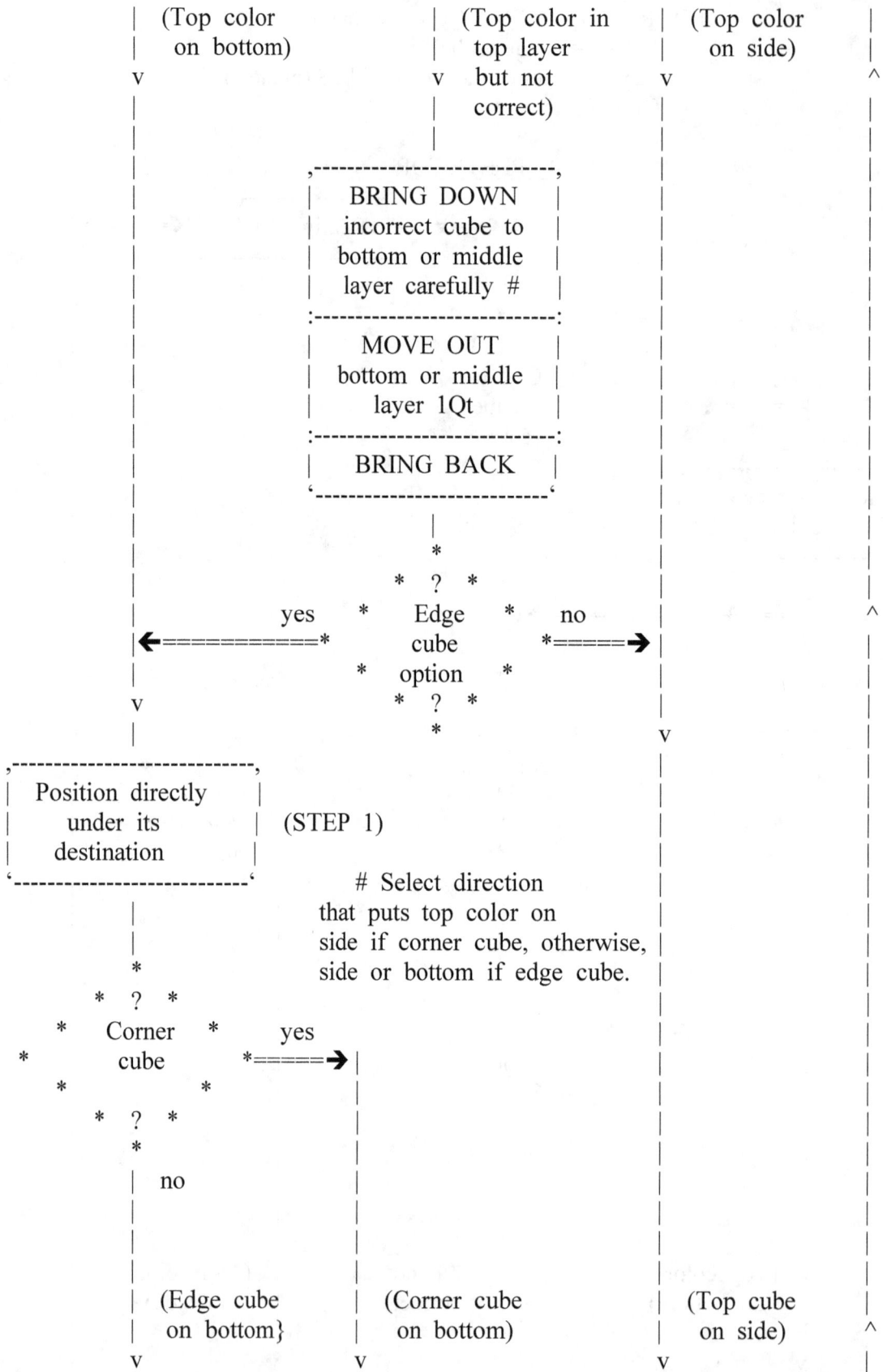

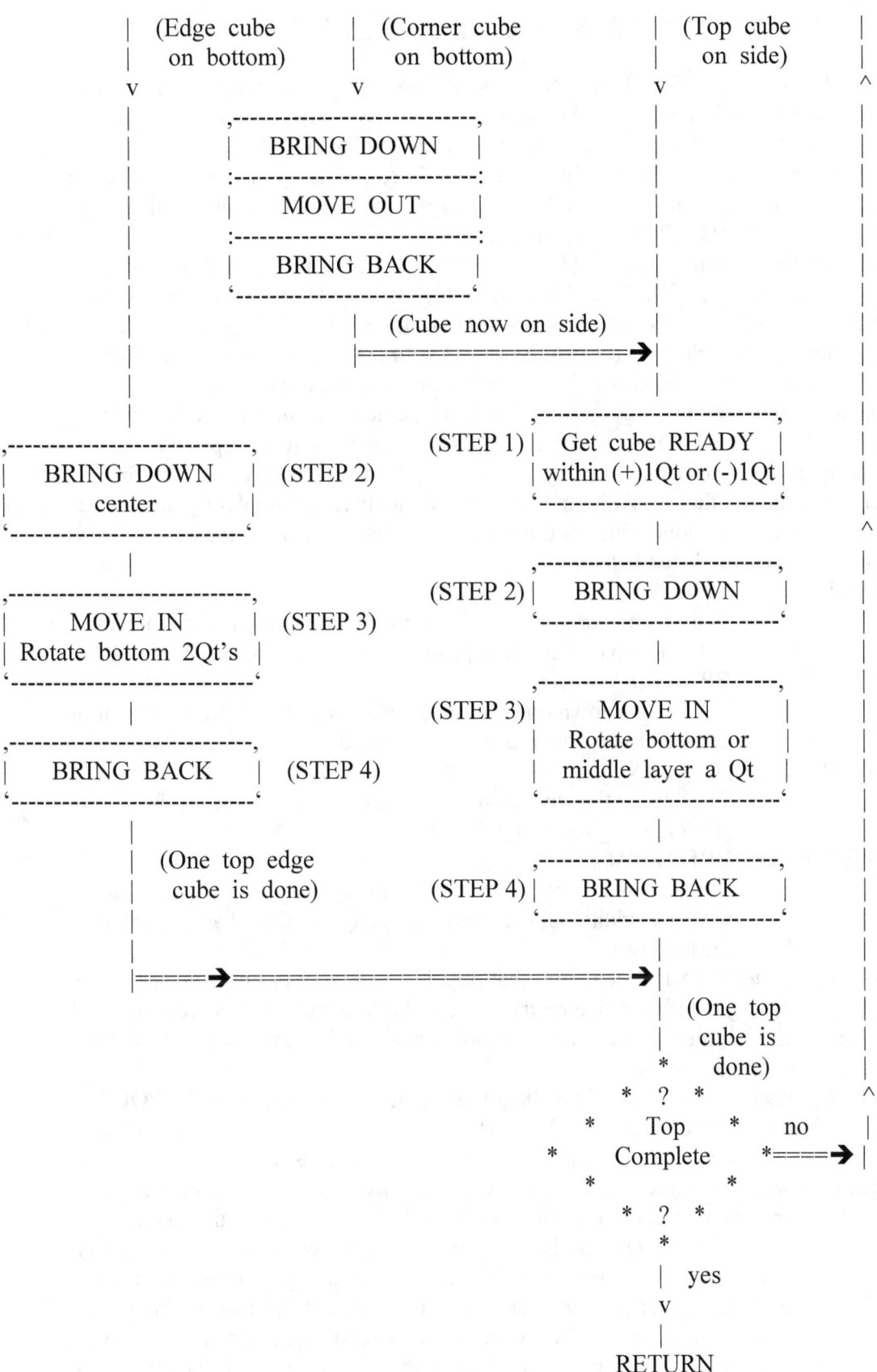

* * * F O U R   S T E P   A P P R O A C H * * *

A detailed review of the FOUR STEP APPROACH used to complete one face of the CUBE is described below, is diagrammed on pages 5, 6, and 7, and is outlined in a CUBIK SUBROUTINE flowchart as shown by Sheets 1S, 2S, and 3S.

This review assumes that the face to be completed is the top face so it is necessary to select a top color and also to eventually select a reference side face color while using the FOUR STEP APPROACH to complete the top.

The challenge of the FOUR STEP APPROACH is that any top (t) color cube ('cublet') that is on either of the four sides (not in the top layer) can be positioned and oriented correctly with four movements (or steps) of the CUBE. A top (t) color EDGE cube on the bottom face can also be positioned and oriented correctly with four steps. However, a top (t) color CORNER cube on the bottom face requires two four-step sequences to position and orient it correctly. A top (t) color cube that is already in the top layer but is not positioned or oriented correctly also requires two four-step sequences. Consequently, a series of FOUR STEP sequences one-by-one correctly positions and orients cubes in the top face until all nine cubes are done thereby completing the top face. (The first cube is already done which is the center cube that determines the color of each face.) The four steps are listed below.

STEP 1:   READY
            This is a preparatory move that rotates the bottom or middle layer
            until the top (t) cube is in proper position for STEPS 2, 3, and 4.
STEP 2:   BRING DOWN
            This brings down one section (or slice) of the top layer by rotating
            the section one-quarter (Q) turn toward you.
STEP 3:   MOVE IN
            This moves the cube made ready in STEP 1 into the top layer
            section brought down in STEP 2.
STEP 4:   BRING BACK
            This restores the top layer section brought down in STEP 2 but
            with one additional cube now correctly positioned and oriented in
            the top layer.

After every FOUR STEP sequence that positions and orients a top cube correctly in the top layer it is a good idea to line up the top and the reference side in order to check that you are positioning the cubes correctly before you select the next cube you want to position and orient in the top layer.

The three pages that follow show diagrams for nine examples using the FOUR STEP APPROACH. Five diagrams (A, B, C, D, E) pertain to the top cube being on one of the four sides. I always do these cubes first. Before starting, as indicated above, you must eventually select two colors so for this review let's assume you have selected green as the top (t) color. The reference side color is selected as soon as either the first corner or edge cube is done. The (*) cubes in the diagram, can be any color, are used to clarify the top section (or slice) that is brought down in STEP 2. Naturally, in order to complete the top, the FOUR STEP sequence is done on any of the four sides that are facing you. Once a cube is correctly positioned and oriented in one diagram it is shown in succeeding diagrams. This might add some confusion to the diagrams but it illustrates how one-by-one the top layer is completed. Next after doing all the top cubes that where on the sides, I do the edge cubes whose top color is facing the bottom, see diagram G. Edge cubes have two colors while corner cubes have three colors. Next I do corner cubes whose top color is facing the bottom, however, they require two FOUR STEP sequences (first to move them to the side, see diagram F; second to move them to the top layer). Finally, I do the top cubes that are in the top layer but are not correct, however, they also require two FOUR STEP sequences (first to move the corner cubes to the side, see diagrams H and J, and to move the edge cubes either to the side or to the bottom face; second to move them back to the top layer).

Diagram A1 assumes that you have started to use the FOUR STEP APPROACH to solve one face but when you started none of the top cubes were correct, all that was correct is the green top center cube. It also assumes that as you were looking over the sides of the CUBE for a corner cube that belongs on the top you located the t,k,r corner cube on the bottom layer which belongs in the x,y,z top corner so you rotated the bottom layer until it is minus a Q turn away thus finishing STEP 1 (the READY step). A2, A3, and A4 show that in STEP 2 you BRING DOWN part of the top layer so that in STEP 3 you rotate the bottom layer to MOVE IN the t,k,r cube in place of the x,y,z cube and when you restore the top in STEP 4 (BRING BACK) the t,k,r cube is now correctly positioned and oriented in the top layer. Now you must select the reference side color so for this review let's select blue as the right (r) face color. This means from now on after every FOUR STEP sequence that places a cube in the top layer you must align the middle layer so that the blue center cube (r) is in alignment with the blue/green corner cube otherwise it is easy to make mistakes and you'll have to do the sequence over again. If the x,y,z cube happens to be a top cube (e.g. t,y,z) don't worry about it, just bump it out with the correct t,k,r cube. The t,y,z cube will be correctly placed by a later sequence.

Diagram B1 shows the t,k,r cube already correctly placed and assumes that you have located the t,e,f corner cube on the bottom layer which belongs in the x,y,z top corner so you rotated the bottom layer so it is plus a Q turn away which finishes STEP 1. Doing STEPS 2, 3, and 4 correctly puts the t.e.f cube in the top corner as shown in diagram B4. If when doing the READY step, STEP 1, the cube is plus a Q turn away instead of minus a Q turn away (or vice-versa), the FOUR STEP sequence will place the cube in the correct corner but it will be incorrectly oriented. After a little practice and a few mistakes, your recognizing correct READY positions will improve. Finding the correct READY positions in STEP 1 is the key in using this FOUR STEP APPROACH.

Diagram C1 shows that the middle layer has been rotated so that the t,k edge cube is minus a Q turn away from replacing the x,y cube in the top layer. STEP 2 (C2) brings down the x,y cube, STEP 3 (C3) replaces the x,y cube with the t,k cube so that STEP 4 (C4) restores the top but with the t,k edge cube correctly placed. The center cubes are labeled with a question mark (?) since the middle layer was rotated during STEP 1, however, after STEP 4 it is important to realign the middle layer before seeking the next cube to be moved to the top layer. An example for the plus Q turn away is shown in diagram E where the t,e edge cube in the middle layer is to replace the x,y cube. Notice that when in the (-) Q turn away situation of diagrams A or C you'll probably use your right hand to BRING DOWN the top layer section in STEP 2 but when in the (+) Q turn away situation of diagrams B or E you'll probably use your left hand.

Diagram D1 shows that the bottom layer has been rotated so that the t,r edge cube is minus a Q turn away from replacing the x,y cube in the top layer. STEP 2 (D2) brings down the x,y cube, STEP 3 (D3) replaces the x,y cube with the t,r cube so that STEP 4 (D4) restores the top but with the t,r edge cube correctly placed. It so happens that getting a bottom layer edge cube in the READY position either a minus Q turn away (as shown in diagram D1) or a plus Q turn away (not shown in a diagram) gives the same result which is not true for three color corner cubes nor for edge cubes in the middle layer.

Diagram E1 shows that the middle layer has been rotated so that the t,e edge cube is plus a Q turn away from replacing the x,y cube in the top layer. STEP 2 (E2) brings down the x,y cube so that STEP 3 (E3) replaces it with the t,e edge cube then STEP 4 (E4) restores the top and now the top layer is two thirds complete.

Diagram F1 shows the case where a corner cube has its top color facing the bottom so the bottom layer is rotated until the t,r,f corner cube is located directly under the top corner where it belongs, which is directly under the x,y,z cube. This is necessary so that STEPS 2, 3, and 4, which move the top color from the bottom face to one of the four side faces, will not mess up any of the top cubes already done. STEP 2 (F2) brings down the x,y,z corner cube. STEP 3 (F3) rotates the bottom layer plus a Q turn which results in the t,r,f corner cube appearing on the left (e) face. STEP 4 (F4) restores the top. Once the t,r,f corner cube is on a side face then using either of the sequences shown in diagrams A or B would place the t,r,f corner cube in its correct top layer corner position. Diagram G1 shows that the t,r,f corner cube has been correctly placed.

Diagrams G1 shows the case where an edge cube has its top color on the bottom face so the bottom layer is rotated until the t,f edge cube is located directly under the top edge where it belongs, which is directly under the x,y cube. STEP 2 (G2) brings down the x,y cube so that in STEP 3 (G3) moving the x,y cube two Q turns results in replacing the x,y cube with the f,t cube and STEP 4 (G4) restores the top but with the f,t edge cube correctly placed.

Diagrams H and J show two cases where a top corner cube is in its correct position but is not oriented correctly. STEPS 2, 3, and 4 for these two cases show how the t,e,k incorrectly oriented corner cube is brought down and dumped off so that the (t) color is on one of the four side faces. If this is not done correctly, the top color will end up being on the bottom face which means that three FOUR STEP sequences are required to place the t,e,k cube properly instead of two FOUR STEP sequences.

The case where a top edge cube is in its correct position but is not oriented correctly is not shown in a diagram. However, as like the corner cubes of diagrams H and J, bring down this edge cube to the bottom layer and dump it off. Now the top color of this edge cube is now on the bottom face but using the sequence in diagram G will reposition and correctly orient this edge cube in the top layer.

TOP LAYER
COMPLETE

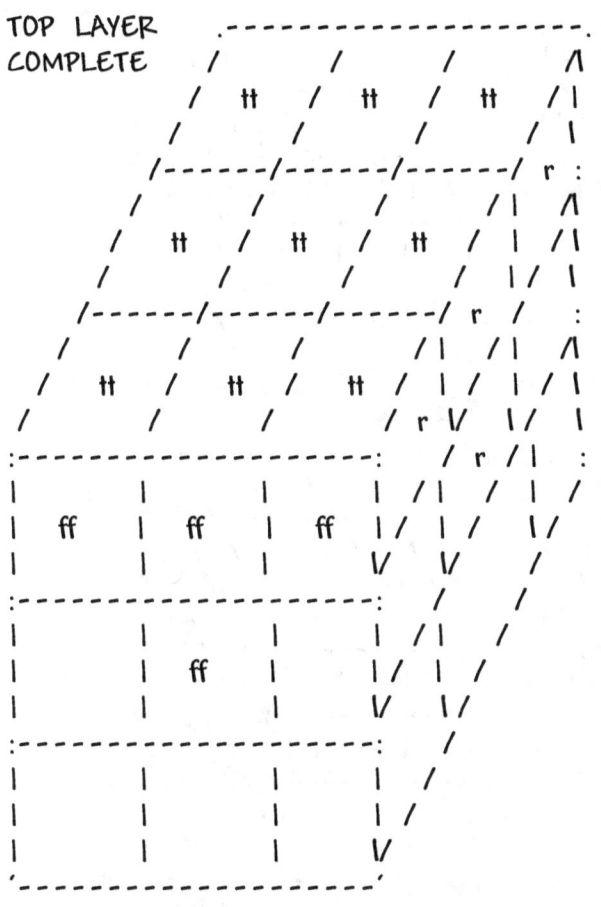

This concludes a rather detailed description of how to solve one face of the CUBE using a FOUR STEP APPROACH. A detailed flow-chart of a FOUR STEP APPROACH appears as a sub-routine in the flowchart section. As shown by the flowchart (after selecting a top color) solving the top face of the CUBE entails first placing at least one edge or corner cube correctly in the top face, then use that cube to select a reference side color. Finally, doing repetitions of the four steps one-by-one places cubes in the top face until the other seven top layer cubes are done as shown in the cubic diagram at the left.

NOTES:

```
A1        .-----------,        B1        .-----------,        C1        .-----------, S
      / * / * / x /|--xyz          / * /  / t /|             / * / x / t /|
     /---/---/---/ y :            /---/---/---/ r :          /---/---/---/ r : T
    /   / t /   /|/|            / * / t /   /|/|          /   / t /   /|/|
    /---/---/---/  / :            /---/---/---/  / :          /---/---/---/  / : E
   /   /   /   /|/|/|            / x /   /   /|/|/|          / t /   /   /|/|/|
  :-----------: / r / :          :-----------: / r / :        :-----------: /?/ : P
  |   |   |   |/|/|/|            | y |   |   |/|/|/|          | f |   |   |/|/|/|
  :-----------: / /             :-----------: / /            :-----------:K/ /    I
  |   | f |   |/|/              |   | f |   |/|/             |   | ? | t |/|/
  :-----------:K/               :-----------:t/              :-----------: /
  |   |   | t |/--tkr            |   |   | e |/--tef          |   |   |   |/
  '-----------'                  '-----------'                '-----------'

A2        .-----------,        B2        .-----------,        C2        .-----------, S
     /   /   /   /|             /   /   / t /|             /   /   /   /|
    /---/---/---/ * :            /---/---/---/ r :          /---/---/---/ * : T
   /   / t /   /|/|            /   / t /   /|/|          /   / t /   /|/|
   /---/---/---/  /* :           /---/---/---/  / :          /---/---/---/ x : E
  /   /   /   /|/|/|            /   /   /   /|/|/|          / t /   /   /|/|/|
  :-----------: / r / x :        :-----------: / r / :        :-----------: /?/ t : P
  |   |   |   |/|/|/|            | * |   |   |/|/|/|          | f |   |   |/|/|/|
  :-----------: / /             :-----------: / /            :-----------:K/ /    2
  |   | f |   |/|/              | * | f |   |/|/             |   | ? | t |/|/
  :-----------:K/               :-----------:t/              :-----------: /
  |   |   | t |/                | x |   | e |/               |   |   |   |/
  '-----------'                  '-----------'                '-----------'

A3        .-----------,        B3        .-----------,        C3        .-----------, S
     /   /   /   /|             /   /   / t /|             /   /   /   /|
    /---/---/---/ * :            /---/---/---/ r :          /---/---/---/ * : T
   /   / t /   /|/|            /   / t /   /|/|          /   / t /   /|/|
   /---/---/---/  /* :           /---/---/---/  / :          /---/---/---/ t : E
  /   /   /   /|/|/|            /   /   /   /|/|/|          / t /   /   /|/|/|
  :-----------: / r / t :        :-----------: / r / :        :-----------: /?/ t : P
  |   |   |   |/|/|/|            | * |   |   |/|/|/|          | f |   |   |/|/|/|
  :-----------: / /             :-----------: / /            :-----------: / /    3
  |   | f |   |/|/              | * | t |   |/|/             |   | ? |   |/|/
  :-----------: /               :-----------: /              :-----------: /
  |   |   |   |/                | t |   |   |/               |   |   |   |/
  '-----------'                  '-----------'                '-----------'

A4        .-----------,        B4        .-----------,        C4        .-----------, S
      / * / * / t /|             / * /   / t /|             / * / t / t /|
     /---/---/---/ r :            /---/---/---/ r :          /---/---/---/ r : T
    /   / t /   /|/|            / * / t /   /|/|          /   / t /   /|/|
    /---/---/---/  / :            /---/---/---/  / :          /---/---/---/  / : E
   /   /   /   /|/|/|            / t /   /   /|/|/|          / t /   /   /|/|/|
  :-----------: / r / :          :-----------: / r / :        :-----------: /?/ : P
  |   |   |   |/|/|/|            | f |   |   |/|/|/|          | f |   |   |/|/|/|
  :-----------: / /             :-----------: / /            :-----------: / /    4
  |   | f |   |/|/              |   | f |   |/|/             |   | ? |   |/|/
  :-----------: / t,k,r          :-----------: / t,e,f        :-----------: / t,k
  |   |   |   |/ (-) Qt          |   |   |   |/ (+) Qt        |   |   |   |/ (-) Qt
  '-----------'  Away            '-----------'  Away          '-----------'  Away
```

```
D1       .-----------,              E1       .-----------,              F1       .-----------, S
      /   / t / t /|                      /  * / t / t /|                      /  t / t / x /|
     /---/---/---/ r :                   /---/---/---/ r :                   /---/---/---/ y : T
    /  * / t / x /|/|                 xy-/ x / t / t /|/|                  /  t / t /   /|/|
   /---/---/---/ y /  :                 /---/---/---/ r /  :              /---/---/---/   /  : E
  / t /   /   /|/|/|                   / t /   /   /|/|/|                /   / t / t /|/|/|
 :-----------: / r /                  :-----------: / r /              :-----------: f / f / r: P
 | f |   |   |/|/|/                   | f |   |   |/|/|/               |   | e | e |/|/|/-c
 :-----------:   /                    :-----------: t/ /               :-----------:   /      |
 |   | f |   |/|/                     |   | ? | e |/|/                 |   | e |   |/|/
 :-----------:   /                    :-----------:   /                :-----------:   /
 |   | (t) |   |/ (t)=t,r             |   |   |   |/                   |   |   |   |/  c=trf
 '-----------'                        '-----------'                    '-----------'

D2       .-----------,              E2       .-----------,              F2       .-----------, S
      /   / t / t /|                      /   / t / t /|                      /   /   / k /|
     /---/---/---/ r :                   /---/---/---/ r :                   /---/---/---/ t : T
    /   / e /   /|/|                    /   / t / t /|/|                   /  t / t /   /|/|
   /---/---/---/ * /  :               /---/---/---/ r /  :               /---/---/---/  / t : E
  / t /   /   /|/|/|                  /   /   /   /|/|/|                 /  t / t / t /|/|/|
 :-----------: / t /                  :-----------: / ? /              :-----------: f / f / x: P
 | f |   |   |/|/|/                   | * |   |   |/|/|/               |   | e | e |/|/|/
 :-----------: / x /                  :-----------: t/ /               :-----------:  / /      2
 |   | f |   |/|/                     | x | ? | e |/|/                 |   | e |   |/|/
 :-----------:   /                    :-----------:   /                :-----------:   /
 |   | (t) |   |/                     | t |   |   |/                   |   |   |   |/
 '-----------'                        '-----------'                    '-----------'

D3       .-----------,              E3       .-----------,              F3       .-----------, S
      /   / t / t /|                      /   / t / t /|                      /   /   / k /|
     /---/---/---/ r :                   /---/---/---/ r :                   /---/---/---/ t : T
    /   / e /   /|/|                    /   / t / t /|/|                   /  t / t /   /|/|
   /---/---/---/ * /  :               /---/---/---/ r /  :               /---/---/---/ / t : E
  / t /   /   /|/|/|                  /   /   /   /|/|/|                 /   / t / t /|/|/|
 :-----------: / t /                  :-----------: / ? /              :-----------: f / f /  : P
 | f |   |   |/|/|/                   | * |   |   |/|/|/               |   | e | e |/|/|/
 :-----------: / (t)/                 :-----------:  / /                :-----------:  / /      3
 |   | f |   |/|/                     | t | ? |   |/|/                 |   | e |   |/|/
 :-----------:   /                    :-----------:   /                :-----------:   /
 |   |   |   |/                       | t |   |   |/                 c-| t |   |   |/  c=trf
 '-----------'                        '-----------'                    '-----------'

D4       .-----------,              E4       .-----------,              F4       .-----------, S
      /   / t / t /|                      /  * / t / t /|                      /  t / t /   /|
     /---/---/---/ r :                   /---/---/---/ r :                   /---/---/---/  : T
    /  * / t / t /|/|                 te-/ t / t / t /|/|                  /  t / t /   /|/|
   /---/---/---/ r /  :               /---/---/---/ r /  :               /---/---/---/ / : E
  / t /   /   /|/|/|                  / t /   /   /|/|/|                 /   / t / t /|/|/|
 :-----------: / r /                  :-----------: / ? /              :-----------: / f / : P
 | f |   |   |/|/|/                   | f |   |   |/|/|/               |   | e | e |/|/|/
 :-----------:  / /                   :-----------:  / /                :-----------:  / /      4
 |   | f |   |/|/                     |   | ? |   |/|/                 |   | e |   |/|/
 :-----------: / t,r                  :-----------: / t,e              :-----------: / t,r,f
 |   |   |   |/ (-) Qt                 |   |   |   |/ (+) Qt            | t |   |   |/ to a
 '-----------' Away                   '-----------' Away               '-----------' side
```

```
G1        .-----------,              H1        ,-----------,              J1        .-----------, S
        / t / t / t /|                       / t / t / t /|                       / t / t / t /|
      /---/---/---/ f:                      /---/---/---/ e:                      /---/---/---/ e : T
      / t / t / x /|/|                      / t / t / t /|/|                      / t / t / t /|/|
    /---/---/---/ y/ :                     /---/---/---/ e/ :                    /---/---/---/ e/  : E
    /   / t / t /|/|/|                    / t / t / e /|/|/|                    / t/ t / K /|/|/|
  :-----------: f/ f/                   :-----------: k/ e/                   :-----------: t/ e/  : P
  |   | e| e |/|/|/                     | K | K | t |/|/|/                    | K | K | e |/|/|/
  :-----------: / f /-ft                :-----------: / /                     :-----------: / /    |
  |   | e |   |/|/                      |   | K |   |/|/                       |   | K |   |/|/
  :-----------: /                       :-----------: /                        :-----------: /
  |   |   |   |/                        |   |   |   |/                          |   |   |   |/
  '-----------'                         '-----------'                           '-----------'
```

```
G2        .-----------,              H2        ,-----------,              J2        ,-----------, S
        / t / t / t /|                       / t / t / t /|                       / t / t /   /|
      /---/---/---/ f:                      /---/---/---/ e:                      /---/---/---/  : T
      /   /   / K /|/|                      / t / t / t /|/|                      / t / t /   /|/|
    /---/---/---/ t/ :                     /---/---/---/ e/ :                    /---/---/---/ /  : E
    /   / t / t /|/|/|                    /   / r /|/|/|                        / t/ t / f /|/|/|
  :-----------: f/ t/  :                 :-----------: t/ e/  :                :-----------: e/ e/  : P
  |   | e| e |/|/|/                     |   |   | K |/|/|/                     | K | K | t |/|/|/
  :-----------: / x /                   :-----------: t/ /                     :-----------: e/ /   2
  |   | e |   |/|/                      |   | K | K |/|/                       |   | K | t |/|/
  :-----------: /                       :-----------: e/                       :-----------: t /--tek
  |   |   |   |/                        |   |   | t |/--tek                     |   |   | K |/
  '-----------'                         '-----------'                           '-----------'
```

```
G3        .-----------,              H3        ,-----------,              J3        ,-----------, S
        / t / t / t /|                       / t / t / t /|                       / t / t /   /|
      /---/---/---/ f:                      /---/---/---/ e:                      /---/---/---/  : T
      /   /   / K /|/|                      / t / t / t /|/|                      / t / t /   /|/|
    /---/---/---/ t/ :                     /---/---/---/ e/ :                    /---/---/---/ /  : E
    /   / t / t /|/|/|                    /   / r /|/|/|                        / t/ t / f /|/|/|
  :-----------: f/ t/  :                 :-----------: t/ e/ t:-tek           :-----------: e/ e/  : P
  |   | e| e |/|/|/                     |   |   | K |/|/|/                     | K | K | t |/|/|/
  :-----------: / t /                   :-----------: t/ /                     :-----------: e/ /   3
  |   | e |   |/|/                      |   | K | K |/|/                       |   | K | t |/|/
  :-----------: /                       :-----------: /                        :-----------: /
  |   |   |   |/                        |   |   |   |/                   tek--| t |   |   |   |/
  '-----------'-->2Qt's                 '-----------'                          '-----------'
```

```
G4        .-----------,              H4        ,-----------,              J4        ,-----------, S
        / t / t / t /|                       / t / t / t /|                       / t / t / t /|
      /---/---/---/ f:                      /---/---/---/ e:                      /---/---/---/ e: T
      / t / t / t /|/|                      / t / t / t /|/|                      / t / t / t /|/|
    /---/---/---/ f/ :                     /---/---/---/ e/ :                    /---/---/---/ e/  : E
    /   / t / t /|/|/|                    / t/ t /   /|/|/|                      / t/ t /   /|/|/|
  :-----------: f/ f/  :                 :-----------: / e / t:-tek           :-----------: / e/  : P
  |   | e| e |/|/|/                     | K | K |   |/|/|/                     | K | K |   |/|/|/
  :-----------: / /                     :-----------: / /                      :-----------: / /    4
  |   | e |   |/|/                      |   | K |   |/|/                       |   | K |   |/|/
  :-----------: / f,t                   :-----------: / t,e,k                  :-----------: / t,e,k
  |   |   |   |/ w (t) on               |   |   |   |/ not           tek--| t |   |   |   |/ not
  '-----------'  Bottom                 '-----------'  Correct               '-----------'  Correct
```

```
P1        .-----------,
        / t / t / t /|
      /---/---/---/ r :
      / t / t / t /|/|
    /---/---/---/ r / r:
   / t / t / t /|/|/|
  :-----------: r / r / r:
  | f | f | f |/|/|/
  :-----------: r / r /
  | f | f | f |/|/
  :-----------: r /
  | f | f | f |/ SOLID
  '-----------'   CUBE
```

```
P3        .-----------,
        / t / b / t /|
      /---/---/---/ r :
      / t / b / t /|/|
    /---/---/---/ r / r:
   / t / b / t /|/|/|
  :-----------: r / r / r:
  | f | k | f |/|/|/
  :-----------: r / r /
  | f | k | f |/|/
  :-----------: r /
  | f | k | f |/ C layer
  '-----------'   2Qt's
```

```
P6        .-----------,
        / t / k / t /|
      /---/---/---/ r :
      / t / k / t /|/|
    /---/---/---/ r / r:
   / t / k / t /|/|/|
  :-----------: r / r / r:
  | f | t | f |/|/|/
  :-----------: r / r /
  | f | t | f |/|/
  :-----------: r /
  | f | t | f |/ C layer
  '-----------'   I Qt
```

```
P7        .-----------,
        / t / k / t /|
      /---/---/---/ r :
      / t / k / t /|/|
    /---/---/---/ r / k:
   / t / k / t /|/|/|
  :-----------: r / b / r:
  | f | t | f |/|/|/
  :-----------: k / r /
  | r | r | r |/|/
  :-----------: r /
  | f | t | f |/ M layer
  '-----------'   I Qt
```

t = top face color
f = front face color
r = right face color
e = left face color
b = bottom face color
k = back face color
Qt = quarter turn

A pattern is inverted by inverting the center layer 'C', see P3, then inverting the vertical layer 'V', see P4, and finally inverting the middle layer 'M', see P5.

```
P4        ,-----------,
        / t / b / t /|
      /---/---/---/ r :
      / b / t / b /|/|
    /---/---/---/ e / r:
   / t / b / t /|/|/|
  :-----------: r / e / r:
  | f | k | f |/|/|/
  :-----------: r / e /
  | f | k | f |/|/
  :-----------: r /
  | f | k | f |/ V layer
  '-----------'   2Qt's
```

```
P8        ,-----------,
        / t / t / t /|
      /---/---/---/ r :
      / t / r / t /|/|
    /---/---/---/ r / k:
   / t / t / t /|/|/|
  :-----------: r / b / r:
  | f | f | f |/|/|/
  :-----------: k / r /
  | r | f | r |/|/
  :-----------: r /
  | f | f | f |/ Return
  '-----------' C layer I Qt
```

```
P2        ,-----------,
        / t / b / t /|
      /---/---/---/ r :
      / b / t / b /|/|
    /---/---/---/ e / e:
   / t / b / t /|/|/|
  :-----------: r / r / r:
  | f | k | f |/|/|/
  :-----------: e / e /
  | k | f | k |/|/
  :-----------: r /
  | f | k | f |/ Inverted
  '-----------'   SOLID
```

```
P5        ,-----------,
        / t / b / t /|
      /---/---/---/ r :
      / b / t / b /|/|
    /---/---/---/ e / e:
   / t / b / t /|/|/|
  :-----------: r / r / r:
  | f | k | f |/|/|/
  :-----------: e / e /
  | k | f | k |/|/
  :-----------: r / 'PONS'
  | f | k | f |/ M layer
  '-----------'   2 Qt's
```

A Scientific American article ,March 1981, calls the inverted solid pattern "The Pons Asinorum" configuration which has been translated Bridge of Asses since it is the easiest pattern to make from a solid CUBE as shown by P2 above. Inverting again, carefully, returns to the solid color of P1.

To make a DOT pattern from a solid CUBE, rotate center layer 'C' I Qt as in P6. Then rotate middle layer 'M' I Qt as in P7. Return center layer 'C' I Qt as in P8 and finally return middle layer 'M' I Qt as in P9 to get the DOT pattern. The DOT pattern has three colors over three faces. This dot pattern phase is called here DOTS1, see DOTS2 in P10.

```
P9        ,-----------,
        / t / t / t /|
      /---/---/---/ r :
      / t / r / t /|/|
    /---/---/---/ r / r:
   / t / t / t /|/|/|
  :-----------: r / f / r:
  | f | f | f |/|/|/
  :-----------: r / r /
  | f | t | f |/|/
  :-----------: r / DOTS1
  | f | f | f |/ Return
  '-----------' M layer IQt
```

```
P10        .-----------,
         / t / t  / t / |
        /---/---/---/ r :
       / t / f / t /|/ |
      /---/---/---/ r / r :
     / t / t / t /|/|/|/|
    :-----------: r / t / r :
    | f | f | f |/|/|/|/
    :-----------: r / r /
    | f | r | f |/|/|/
    :-----------: r /
    | f | f | f |/ DOTS2
    '-----------'
```

```
P11A       .-----------,
         / t / b / t / |
        /---/---/---/ r :
       / b / f / b /|/ |
      /---/---/---/ e / e :
     / t / b / t /|/|/|/|
    :-----------: r / t / r :
    | f | k | f |/|/|/|/
    :-----------: e / e /
    | k | r | k |/|/|/
    :-----------: r /
    | f | k | f |/ DOTS2
    '-----------' inverted
```

```
P12        .-----------,
         / t / t  / t / |
        /---/---/---/ r :
       / t / t / t /|/ |
      /---/---/---/ r / r :
     / t / t / t /|/|/|/|
    :-----------: r / e / r :
    | f | f | f |/|/|/|/
    :-----------: r / r /
    | f | k | f |/|/|/
    :-----------: r /
    | f | f | f |/ IMAGE
    '-----------'  DOTS
```

```
P13        .-----------,
         / t / b / t / |
        /---/---/---/ r :
       / b / t / b /|/ |
      /---/---/---/ e / e :
     / t / b / t /|/|/|/|
    :-----------: r / e / r :
    | f | k | f |/|/|/|/
    :-----------: e / e /
    | k | k | k |/|/|/
    :-----------: r / IMAGE
    | f | k | f |/ DOTS
    '-----------' inverted
```

Starting with the dot pattern as shown in P9, repeating P6, P7, P8, and P9 results in the other phase of the dot pattern as shown in P10 (same three colors over three faces) and is called here DOTS2. The dot pattern is cyclic in three, so repeating P6, P7, P8, and P9 starting with the dot pattern in P10 results in a solid CUBE again. P11A shows the P10 dot pattern inverted. Starting with P10, repeat P3, P4, and P5 in order to get P11A, DOTS2 inverted. The inverted dot pattern has all six colors over three faces. (Inverting patterns is cyclic in two.) P11B shows the DOTS1 pattern inverted. Inverting either P11A or P11B again returns to the dot patterns of P9 and P10. Note: the center cubes stay put after inverting patterns.

To recover from a dot pattern, make an 'H' pattern with the first Q turn on either of the faces 't', 'r', or 'f' then you should be able to return to a solid color.

```
P11B       .-----------,
         / t / b / t / |
        /---/---/---/ r :
       / b / r / b /|/ |
      /---/---/---/ e / e :
     / t / b / t /|/|/|/|
    :-----------: r / f / r :
    | f | k | f |/|/|/|/
    :-----------: e / e /
    | k | t | k |/|/|/
    :-----------: r /
    | f | k | f |/ DOTS1
    '-----------' inverted
```

If you don't make an 'H' with the first Qt, then you'll recover to an IMAGE DOT pattern as shown in P12. (To get from P10 to P12, rotate the center slice 'C' I Qt away from you then rotate the middle layer 'M' I Qt to the right so the three 't' colors are in a row on the back face of the CUBE, finally recover the top and then recover the middle layer; similarly to get from P9 to P12, rotate the vertical slice 'V' I Qt to the left and then rotate the middle layer 'M' I Qt to the left so the three 't' colors are in a row on the left face of the CUBE, finally recover the top and then recover the middle layer). Inverting the IMAGE DOT pattern results in two crosses (+) and one (X) pattern as shown in P13. To recover from IMAGE DOT pattern, move the center layer 'C' I Qt and then recover to a DOT pattern from which you can recover to a solid CUBE again. Note: there are many phases of the DOT and IMAGE DOT patterns, e.g. if you move the middle layer 'M' in the opposite direction after bringing down the center slice 'C' and then recover you end up with a DOT pattern of three colors over three faces but the faces are the top, front, and left instead of the top, front, and right of P9.

So far we have learned to invert a pattern by moving the center, vertical, and middle layers two Q turns. By moving the center and middle layers one Q turn we were able to make DOT patterns and an IMAGE DOT pattern.

As you become more proficient in pattern making, remember that with four additional moves you can add a DOT pattern to any other pattern (actually, four moves involves eight twists of the CUBE).

This next little series of pattern sequences results in making an IMAGE CROSS and an IMAGE X pattern, see P18 and P20.

```
P14A       .-----------,
         / t / t / t / |
        /---/---/---/ e :
       / t / t / t /|/ |
      /---/---/---/ e / e :
     / t / t / t /|/|/ |
    :-----------: e / r / e :
    | ĸ | f | ĸ |/|/|/ |
    :-----------: e / e /
    | ĸ | f | ĸ |/|/
    :-----------: e /
    | ĸ | f | ĸ |/
    '-----------'  DOT/BAR1
```

```
P16        .-----------,
         / t / t / t / |
        /---/---/---/ r :
       / t / t / t /|/ |
      /---/---/---/ r / r :
     / t / t / t /|/|/ |
    :-----------: r / r / r :
    | f | ĸ | f |/|/|/ |
    :-----------: r / r /
    | f | f | f |/|/
    :-----------: r /
    | f | ĸ | f |/  SOLID/
    '-----------'  SOLID/H
```

```
P18        .-----------,
         / t / t / t / |
        /---/---/---/ e :
       / t / t / t /|/ |
      /---/---/---/ r / r :
     / t / t / t /|/|/ |
    :-----------: e / r / e :
    | ĸ | f | ĸ |/|/|/ |
    :-----------: r / r /
    | f | f | f |/|/
    :-----------: e /
    | ĸ | f | ĸ |/  IMAGE
    '-----------'  CROSS
```

```
P20        .-----------,
         / t / t / t / |
        /---/---/---/ r :
       / t / t / t /|/ |
      /---/---/---/ e / e :
     / t / t / t /|/|/ |
    :-----------: r / r / r :
    | f | ĸ | f |/|/|/ |
    :-----------: e / e /
    | ĸ | f | ĸ |/|/
    :-----------: r /
    | f | ĸ | f |/
    '-----------'  IMAGE X
```

```
P14B       .-----------,
         / t / t / t / |
        /---/---/---/ e :
       / t / t / t /|/ |
      /---/---/---/ r / e :
     / t / t / t /|/|/ |
    :-----------: e / r / e :
    | ĸ | ĸ | ĸ |/|/|/ |
    :-----------: e / r /
    | ĸ | f | ĸ |/|/
    :-----------: e /
    | ĸ | ĸ | ĸ |/
    '-----------'  DOT/BAR2
```

One can go directly from a solid
CUBE to an IMAGE DOT pattern
by first inverting the center layer
(as in P3), second, rotate middle
layer 1 Qt to the left, recover the
top and then recover the middle
layer. You should now have an
IMAGE DOT pattern.

   Let's call this last sequence
the *IMAGE DOT Sequence*.
Doing the sequence again returns
to a solid CUBE.

   To get a different set of
patterns from a solid CUBE,
first invert the center (as in P3),
then rotate the top layer + 1 Qt and
the bottom layer – 1 Qt; same result
if top – 1 Qt and bottom + 1 Qt.
Recover the top (top is a solid
color), then recover the sides
which results in a DOT/BAR1
pattern of P14A. If the vertical
slice is inverted instead of the
center slice, then DOT/BAR2
results, another phase of the
DOT/BAR patterns. Let's label
this the *DOT/BAR Sequence*.
Inverting the DOT/BAR1 pattern
results in a X. BAR, CROSS
pattern, see P15. Starting with
P14A and doing an IMAGE DOT
Sequence results in P16 SOLID/
SOLID/H pattern. Inverting P16
results in a X, X, H pattern of P17.
Repeat IMAGE DOT Sequence on
P16 returns to P14A. Repeating
DOT/BAR Sequence on P16
returns to IMAGE DOT pattern.
Starting with P14A and doing a

```
P15        .-----------,
         / t / b / t / |
        /---/---/---/ e :
       / b / t / b /|/ |
      /---/---/---/ r / r :
     / t / b / t /|/|/ |
    :-----------: e / r / e :
    | ĸ | ĸ | ĸ |/|/|/ |
    :-----------: r / r /
    | f | f | f |/|/
    :-----------: e /
    | ĸ | ĸ | ĸ |/
    '-----------'  X,BAR,CROSS
```

```
P17        .-----------,
         / t / b / t / |
        /---/---/---/ r :
       / b / t / b /|/ |
      /---/---/---/ e / e :
     / t / b / t /|/|/ |
    :-----------: r / r / r :
    | f | f | f |/|/|/ |
    :-----------: e / e /
    | ĸ | f | ĸ |/|/
    :-----------: r /
    | f | f | f |/
    '-----------'  X,X,H
```

```
P19        .-----------,
         / t / b / t / |
        /---/---/---/ e :
       / b / t / b /|/ |
      /---/---/---/ e / e :
     / t / b / t /|/|/ |
    :-----------: e / r / e :
    | ĸ | ĸ | ĸ |/|/|/ |
    :-----------: e / e /
    | ĸ | f | ĸ |/|/
    :-----------: e /
    | ĸ | ĸ | ĸ |/
    '-----------'  DOT/DOT/X
```

```
P21        .-----------,
         / t / b / t / |
        /---/---/---/ r :
       / b / t / b /|/ |
      /---/---/---/ r / r :
     / t / b / t /|/|/ |
    :-----------: r / r / r :
    | f | f | f |/|/|/ |
    :-----------: r / r /
    | f | f | f |/|/
    :-----------: r /
    | f | f | f |/  SOLID/
    '-----------'  SOLID/X
```

IMAGE DOT Sequence on the right face (the DOT face) also results in a SOLID/SOLID/H pattern. However, doing a DOT/BAR Sequence on the right face results in the IMAGE CROSS of P18 but doing a DOT/BAR Sequence on the front face returns to a solid CUBE. Inverting P18 gives DOT/DOT/X pattern of P19. Doing an IMAGE DOT Sequence on P18 results in a IMAGE X pattern of P20. Inverting P20 results in SOLID/SOLID/X pattern of P21. Repeating an IMAGE DOT Sequence on P20 returns to the IMAGE CROSS pattern. Doing a DOT/BAR Sequence on P18 returns to P14A but it must be on the right or left face, (r) or (e) center cubes, or it returns to P14B instead. *NOTE: for these patterns above the views of the CUBE always show the t, f, r center cubes.*

I always made the four patterns on the left (below), P22, P24, P26, and P28, starting from a scrambled CUBE. P22 is called PLUMMER-CROSS by the Scientific American article. It is done with three colors over three faces. Inverting P22 results in a pattern that looks very similar to P11, INVERTED DOT pattern.

P24 is called the CHRISTMAN-CROSS configuration by the Scientific American article. It is done with four colors over three faces. Inverting P24 results in a pattern that contains an IMAGE DOT pattern which is shown in P25. Since pattern P26 has four colors over three faces, I call it the CHRIST-MAN HHX pattern. Inverting P26 results in pattern P27 which contains two colorful X patterns. I call P28 the PLUMMER-XXX pattern since it has three colors over three faces. Inverting P28 also results in three X patterns but has six colors over three faces as shown in pattern P29.

Taylor/Ryland shows most of these patterns made from a solid CUBE. If you do their algorithms twice the patterns return to a solid CUBE, therefore, this group of patterns is cyclic in two. All of the MASON type patterns shown on the next two pages are cyclic in three so using the Taylor/Ryland algorithms to make them from a solid CUBE results in making one phase of the pattern. Repeating the algorithm makes another phase. *(Two views are shown for the six MASON patterns which illustrates how a MASON pattern on the t, r, f faces is repeated, 'shows up', on the b, e, k faces.)* I made this MASON group of patterns starting from a scrambled CUBE. They all have the

```
P22      .-----------,
        / f / t / f /|
       /---/---/---/ t :
      / t / t / t /|/|
     /---/---/---/ r / r :
    / f / t / f /|/|/|
   :-----------: t / r / t :
   | r | f | r |/|/|/
   :-----------: r / r /
   | f | f | f |/|/
   :-----------: t /
   | r | f | r |/ PLUMMER
   '-----------'   CROSS
```

```
P24      .-----------,
        / b / t / b /|
       /---/---/---/ f :
      / t / t / t /|/|
     /---/---/---/ r / r :
    / b / t / b /|/|/|
   :-----------: f / r / f :
   | r | f | r |/|/|/
   :-----------: r / r /
   | f | f | f |/|/
   :-----------: f /
   | r | f | r |/ CHRIST-
   '-----------'MAN CROSS
```

```
P26      .-----------,
        / t / b / t /|
       /---/---/---/ r :
      / b / t / b /|/|
     /---/---/---/ f / r :
    / t / b / t /|/|/|
   :-----------: r / r / r :
   | f | r | f |/|/|/
   :-----------: r / f /
   | f | f | f |/|/
   :-----------: r /
   | f | r | f |/ CHRIST-
   '-----------'MAN 'HHX'
```

```
P23      .-----------,
        / f / b / f /|
       /---/---/---/ t :
      / b / t / b /|/|
     /---/---/---/ e / e :
    / f / b / f /|/|/|
   :-----------: t / r / t :
   | r | k | r |/|/|/
   :-----------: e / e /
   | k | f | k |/|/
   :-----------: t /
   | r | k | r |/ P22
   '-----------' Inverted
```

```
P25      .-----------,
        / b / b / b /|
       /---/---/---/ f :
      / b / t / b /|/|
     /---/---/---/ e / e :
    / b / b / b /|/|/|
   :-----------: f / r / f :
   | r | k | r |/|/|/
   :-----------: e / e /
   | k | f | k |/|/
   :-----------: f /
   | r | k | r |/ P24
   '-----------' Inverted
```

```
P27      .-----------,
        / t / t / t /|
       /---/---/---/ r :
      / t / t / t /|/|
     /---/---/---/ k / e :
    / t / t / t /|/|/|
   :-----------: r / r / r :
   | f | e | f |/|/|/
   :-----------: e / k /
   | k | f | k |/|/
   :-----------: r /
   | f | e | f |/ P26
   '-----------' Inverted
```

```
P28     .-----------,
       / t / f / t /|
      /---/---/---/ r :
     / f / t / f /|/|
    /---/---/---/ t / t :
   / t / f / t /|/|/|
  :-----------: r / r / r :
  | f | r | f |/|/|/|
  :-----------: t / t /
  | r | f | r |/|/
  :-----------: r /
  | f | r | f |/ PLUMMER
  '-----------' 'XXX'
P30A    .-----------,
       / t / t / t /|
      /---/---/---/ r :
     / t / t / t /|/|
    /---/---/---/ r / r :
   / t / t / f /|/|/|
  :-----------: t / r / r :
  | f | f | r |/|/|/|
  :-----------: r / r /
  | f | f | f |/|/
  :-----------: r /
  | f | f | f |/ BIT
  '-----------' MASON 1
P30B    .-----------,
       / t / t / t /|
      /---/---/---/ r :
     / t / t / t /|/|
    /---/---/---/ r / r :
   / t / t / r /|/|/|
  :-----------: f / r / r :
  | f | f | t |/|/|/|
  :-----------: r / r /
  | f | f | f |/|/
  :-----------: r /
  | f | f | f |/ BIT
  '-----------' MASON 2
P31A    .-----------,
       / f / f / f /|
      /---/---/---/ t :
     / f / t / t /|/|
    /---/---/---/ r / t :
   / f / t / t /|/|/|
  :-----------: r / r / t :
  | r | f | f |/|/|/|
  :-----------: r / t /
  | r | f | f |/|/
  :-----------: t /
  | r | r | r |/ GIANT
  '-----------' MASON 1
```

same three colors over three faces and all six faces are diagrammed.

Patterns P30A and P30B are two phases of the pattern I call the BIT MASON pattern. (It is on the front cover of the March '81 issue of Scientific American.) The b, e, k "DOT" cube is shown diagonally across from the t, r, f "DOT" cube, however, this b, e, k "DOT" orientation can appear on the b, e, f or b, r, k or b, r, f corner instead.

(FOR THESE SIX MASON PATTERNS, ON THE LEFT ARE THE t, r, f VIEWS, ON THE RIGHT ARE THE b, e, k VIEWS.)

Patterns P31A and P31B are two phases of the GIANT MASON pattern. Patterns P31B1 and P31B2 are two phases of the GIANT BIT MASON pattern which can be made by adding patterns P30A or P30B to pattern P31B using Taylor/Ryland's algorithm, otherwise make from a scrambled CUBE. Pattern P31B1 appears in the Scientific American article and was named "The Giant Mason with Cherries Position". Similarly patterns P30A and P30B can be added to P31A to form two additional GIANT BIT MASON patterns (not shown). Patterns P31B1 and P31B2 can also be called a RING AROUND A CORNER pattern using six cubes (three per view).

```
P29     .-----------,
       / t / k / t /|
      /---/---/---/ r :
     / k / t / k /|/|
    /---/---/---/ b / b :
   / t / k / t /|/|/|
  :-----------: r / r / r :
  | f | e | f |/|/|/|
  :-----------: b / b /
  | e | f | e |/|/
  :-----------: r /
  | f | e | f |/ P28
  '-----------' Inverted
P30A'   .-----------,
       / b / b / b /|
      /---/---/---/ k :
     / b / b / b /|/|
    /---/---/---/ k / k :
   / b / b / k /|/|/|
  :-----------: e / k / k :
  | e | e | b |/|/|/|
  :-----------: k / k /
  | e | e | e |/|/
  :-----------: k /
  | e | e | e |/ (b, e, k
  '-----------' view)
P30B'   .-----------,
       / b / b / b /|
      /---/---/---/ k :
     / b / b / b /|/|
    /---/---/---/ k / k :
   / b / b / e /|/|/|
  :-----------: b / k / k :
  | e | e | k |/|/|/|
  :-----------: k / k /
  | e | e | e |/|/
  :-----------: k /
  | e | e | e |/ (b, e, k
  '-----------' view)
P31A'   .-----------,
       / k / k / k /|
      /---/---/---/ e :
     / k / b / b /|/|
    /---/---/---/ k / e :
   / k / b / b /|/|/|
  :-----------: k / k / e :
  | b | e | e |/|/|/|
  :-----------: k / e /
  | b | e | e |/|/
  :-----------: e /
  | b | b | b |/ (b, e, k
  '-----------' view)
```

```
P31B        .----------,
           / r / r / r /|
          /---/---/---/ f :
         / r / t / t /|/|
        /---/---/---/ r / f :
       / r / t / t /|/|/|
      :----------: r / r / f :
      | t | f | f |/|/|/|
      :----------: r / f /
      | t | f | f |/|/|
      :----------: f /
      | t | t | t |/ GIANT
      '----------'  MASON 2
```

```
P31B'       .----------,
           / e / e / e /|
          /---/---/---/ b :
         / e / b / b /|/|
        /---/---/---/ k / b :
       / e / b / b /|/|/|
      :----------: k / k / b :
      | k | e | e |/|/|/|
      :----------: k / b /
      | k | e | e |/|/|
      :----------: b /
      | k | k | k |/ (b, e, k
      '----------'   view)
```

```
P31B1       .----------,
           / r / r / r /|
          /---/---/---/ f :
         / r / t / t /|/|
        /---/---/---/ r / f :
       / r / t / f /|/|/|
      :----------: t / r / f :
      | t | f | r |/|/|/|
      :----------: r / f /
      | t | f | f |/|/|
      :----------: f /
      | t | t | t |/ GIANT BIT
      '----------'  MASON 2A (cherries)
```

```
P31B1'      ,----------,
           / e / e / e /|
          /---/---/---/ b :
         / e / b / b /|/|
        /---/---/---/ k / b :
       / e / b / k /|/|/|
      :----------: e / k / b :
      | k | e | b |/|/|/|
      :----------: k / b /
      | k | e | e |/|/|
      :----------: b /
      | k | k | k |/ (b, e, k
      '----------'   view)
```

```
P31B2       .----------,
           / r / r / r /|
          /---/---/---/ f :
         / r / t / t /|/|
        /---/---/---/ r / f :
       / r / t / r /|/|/|
      :----------: f / r / f :
      | t | f | t |/|/|/|
      :----------: r / f /
      | t | f | f |/|/|
      :----------: f /
      | t | t | t |/ GIANT BIT
      '----------'  MASON 2B
```

```
P31B2'      ,----------,
           / e / e / e /|
          /---/---/---/ b :
         / e / b / b /|/|
        /---/---/---/ k / b :
       / e / b / e /|/|/|
      :----------: b / k / b :
      | k | e | k |/|/|/|
      :----------: k / b /
      | k | e | e |/|/|
      :----------: b /
      | k | k | k |/ (b, e, k
      '----------'   view)
```

The "DOT" pattern has two phases to it as shown before in patterns P9 and in P10. The six sets of pattern diagrams on the next two pages have the 'DOT' pattern added to them. The patterns on the left side are the basic patterns. Adding a DOT sequence (P6, P7, P8, P9) with eight twists of the CUBE results in the center pattern. Adding the same DOT sequence to the center patterns results in the patterns on the right side. Adding the same DOT sequence to the patterns on the right side returns to the basic pattern on the left side. (HERE ONLY ONE VIEW IS SHOWN.)

PATTERNS

```
P30AA    .------------,          P30AA1   .------------,          P30AA2   .------------,
        / † / † / † /|                   / † / † / † /|                   / † / † / † /|
       /---/---/---/ r :                /---/---/---/ r :                /---/---/---/ r :
      / † / † / † /|/|               / † / r / † /|/|                 / † / f / † /|/|
     /---/---/---/ r / r :           /---/---/---/ r / r :             /---/---/---/ r / r :
    / † / † / f /|/|/|             / † / † / f /|/|/|               / † / † / f /|/|/|
   :-----------: † / r / r :        :-----------: † / f / r :          :-----------: † / † / r :
   | f | f | r |/|/|/           | f | f | r |/|/|/               | f | f | r |/|/|/
   :-----------: r / r /           :-----------: r / r /              :-----------: r / r /
   | f | f | f |/|/              | f | † | f |/|/                 | f | r | f |/|/
   :-----------: r /              :-----------: r /                 :-----------: r /
   | f | f | f | f |/ BIT          | f | f | f | f |/ DOT BIT       | f | f | f | f |/ DOT BIT
   '-----------' MASON 1          '-----------' Different          '-----------' Same

P30BB    .------------,          P30BB1   .------------,          P30BB2   .------------,
        / † / † / † /|                   / † / † / † /|                   / † / † / † /|
       /---/---/---/ r :                /---/---/---/ r :                /---/---/---/ r :
      / † / † / † /|/|               / † / r / † /|/|                 / † / f / † /|/|
     /---/---/---/ r / r :           /---/---/---/ r / r :             /---/---/---/ r / r :
    / † / † / r /|/|/|             / † / † / r /|/|/|               / † / † / r /|/|/|
   :-----------: f / r / r :        :-----------: f / f / r :          :-----------: f / † / r :
   | f | f | † |/|/|/           | f | f | † |/|/|/               | f | f | † |/|/|/
   :-----------: r / r /           :-----------: r / r /              :-----------: r / r /
   | f | f | f |/|/              | f | † | f |/|/                 | f | r | f |/|/
   :-----------: r /              :-----------: r /                 :-----------: r /
   | f | f | f | f |/ BIT          | f | f | f | f |/ DOT BIT       | f | f | f | f |/ DOT BIT
   '-----------' MASON 2          '-----------' Same                '-----------' Different

P31AA    .------------,          P31AA1   .------------,          P31AA2   .------------,
        / f / f / f /|                   / f / f / f /|                   / f / f / f /|
       /---/---/---/ † :                /---/---/---/ † :                /---/---/---/ † :
      / f / † / † /|/|               / f / r / † /|/|                 / f / f / † /|/|
     /---/---/---/ r / † :           /---/---/---/ r / † :             /---/---/---/ r / † :
    / f / † / † /|/|/|             / f / † / † /|/|/|               / f / † / † /|/|/|
   :-----------: r / r / †          :-----------: r / f / †           :-----------: r / † / †
   | r | f | f |/|/|/           | r | f | f |/|/|/               | r | f | f |/|/|/
   :-----------: r / † /           :-----------: r / † /              :-----------: r / † /
   | r | f | f |/|/              | r | † | f |/|/                 | r | r | f |/|/
   :-----------: † /              :-----------: † /                 :-----------: † /
   | r | r | r | r |/ GIANT        | r | r | r | r |/ DOT           | r | r | r | r |/ FOUR-BIT
   '-----------' MASON 1          '-----------' MASON 1            '-----------' MASON 1

P31BB    .------------,          P31BB1   .------------,          P31BB2   .------------,
        / r / r / r /|                   / r / r / r /|                   / r / r / r /|
       /---/---/---/ f :                /---/---/---/ f :                /---/---/---/ f :
      / r / † / † /|/|               / r / r / † /|/|                 / r / f / † /|/|
     /---/---/---/ r / f :           /---/---/---/ r / f :             /---/---/---/ r / f :
    / r / † / † /|/|/|             / r / † / † /|/|/|               / r / † / † /|/|/|
   :-----------: r / r / f          :-----------: r / f / f           :-----------: r / † / f
   | † | f | f |/|/|/           | † | f | f |/|/|/               | † | f | f |/|/|/
   :-----------: r / f /           :-----------: r / f /              :-----------: r / f /
   | † | f | f |/|/              | † | † | f |/|/                 | † | r | f |/|/
   :-----------: f /              :-----------: f /                 :-----------: f /
   | † | † | † | † |/ GIANT        | † | † | † | † |/ FOUR-BIT      | † | † | † | † |/ DOT
   '-----------' MASON 2          '-----------' MASON 2            '-----------' MASON 2
```

```
P3IBBI'  .- - - - - - - - - -,          P3IBBI" .- - - - - - - - - -,          P3IBBI"' .- - - - - - - - - -,
         / r / r / r /|                         / r / r / r /|                         / r / r / r /|
        /---/---/---/ f:                        /---/---/---/ f:                        /---/---/---/ f:
       / r / t / t /|/|                        / r / r / t /|/|                        / r / f / t /|/|
      /---/---/---/ r/ f:                      /---/---/---/ r/ f:                      /---/---/---/ r/ f:
     / r / t / f /|/|/|                       / r / t / f /|/|/|                       / r / t / f /|/|/|
    :- - - - - - - -: t/ r/ f:               :- - - - - - - -: t/ f/ f:               :- - - - - - - -: t/ t/ f:
    | t  | f  | r |/|/|/|                     | t  | f  | r |/|/|/|                     | t  | f | r |/|/|/|
    :- - - - - - - -: r/ f/                  :- - - - - - - -: r/ f/                  :- - - - - - - -: r/ f/
    | t  | f  | f |/|/|                       | t  | t | f |/|/|                       | t  | r | f |/|/|
    :- - - - - - - -: f/                     :- - - - - - - -: f/                     :- - - - - - - -: f/
    | t  | t  | t  | |/ GIANT BIT            | t  | t | t  | |/ RING                  | t  | t | t  | |/ GIANT BIT
    '- - - - - - - -'  MASON 2A             '- - - - - - - -' AROUND BIT             '- - - - - - - -' DOT MASON I

P3IBB2'  .- - - - - - - - - -,          P3IBB2" .- - - - - - - - - -,          P3IBB2"' .- - - - - - - - - -,
         / r / r / r /|                         / r / r / r /|                         / r / r / r /|
        /---/---/---/ f:                        /---/---/---/ f:                        /---/---/---/ f:
       / r / t / t /|/|                        / r / r / t /|/|                        / r / f / t /|/|
      /---/---/---/ r/ f:                      /---/---/---/ r/ f:                      /---/---/---/ r/ f:
     / r / t / r /|/|/|                       / r / t / r /|/|/|                       / r / t / r /|/|/|
    :- - - - - - - -: f/ r/ f:               :- - - - - - - -: f/ f/ f:               :- - - - - - - -: f/ t/ f:
    | t  | f  | t |/|/|/|                     | t  | f | t |/|/|/|                     | t  | f | t |/|/|/|
    :- - - - - - - -: r/ f/                  :- - - - - - - -: r/ f/                  :- - - - - - - -: r/ f/
    | t  | f  | f |/|/|                       | t  | t | f |/|/|                       | t  | r | f |/|/|
    :- - - - - - - -: f/                     :- - - - - - - -: f/ RING                :- - - - - - - -: f/
    | t  | t  | t  | |/ GIANT BIT            | t  | t | t  | |/ AROUND                | t  | t | t  | |/ GIANT BIT
    '- - - - - - - -'  MASON 2B             '- - - - - - - -'  CORNER                '- - - - - - - -' DOT MASON 2


NOTES: .- - - - - - - - - -,         .- - - - - - - - - -,           .- - - - - - - - - -,
       /  /  /  /|                   /  /  /  /|                      /  /  /  /|
      /---/---/---/ :                /---/---/---/ :                  /---/---/---/ :
     /  /  /  /|/|                  /  /  /  /|/|                    /  /  /  /|/|
    /---/---/---/ / :              /---/---/---/ / :                /---/---/---/ / :
   /  /  /  /|/|/|               /  /  /  /|/|/|                  /  /  /  /|/|/|
  :- - - - - - - -: / / :        :- - - - - - - -: / / :          :- - - - - - - -: / / :
  |   |   |   |/|/|/|            |   |   |   |/|/|/|              |   |   |   |/|/|/|
  :- - - - - - - -: / /          :- - - - - - - -: / /            :- - - - - - - -: / /
  |   |   |   |/|/|             |   |   |   |/|/|                |   |   |   |/|/|
  :- - - - - - - -: /            :- - - - - - - -: /              :- - - - - - - -: /
  |   |   |   | |/               |   |   |   | |/                 |   |   |   | |/
  '- - - - - - - -'              '- - - - - - - -'                '- - - - - - - -'
```

```
PA1      .-----------,
        / b / b / b /|
       /---/---/---/ f:
      / b / t / b /|/|
     /---/---/---/ f/ f:
    / b / b / b /|/|/|
   :-----------: f/ r/ f:
   | r | r | r |/|/|/
   :-----------: f/ f/
   | r | f | r |/|/
   :-----------: f/
   | r | r | r |/ CHRIST-
   '-----------' MAN DOTS

PA3      .-----------,
        / b / b / b /|
       /---/---/---/ f:
      / b / r / b /|/|
     /---/---/---/ f/ f:
    / b / b / b /|/|/|
   :-----------: f/ f/ f:
   | r | r | r |/|/|/
   :-----------: f/ f/
   | r | t | r |/|/
   :-----------: f/
   | r | r | r |/ PA1 (DOT
   '-----------' Sequence-1)

PA5      .-----------,
        / t / b / t /|
       /---/---/---/ r:
      / b / t / b /|/|
     /---/---/---/ f/ f:
    / t / b / t /|/|/|
   :-----------: r/ r/ r:
   | f | r | f |/|/|/
   :-----------: f/ f/
   | r | f | r |/|/
   :-----------: r/
   | f | r | f |/ CHRIST-
   '-----------' MAN X

PA7      .-----------,
        / t / t / t /|
       /---/---/---/ r:
      / b / t / b /|/|
     /---/---/---/ r/ f:
    / t / t / t /|/|/|
   :-----------: r/ r/ r:
   | f | f | f |/|/|/
   :-----------: f/ r/
   | r | f | r |/|/
   :-----------: r/
   | f | f | f |/ CHRIST-
   '-----------' MAN H
```

By taking the CUBE apart and interchanging two edge cubes, these patterns are made from an altered-scrambled CUBE. Since PA1 has four colors over three faces, I call it the CHRISTMAN DOTS pattern. Inverting PA1 results in PA2. PA3 is doing a DOT sequence on PA1. PA4 is doing a DOT sequence on PA3. Doing a DOT sequence on PA4 returns to PA1. Since PA5 has four colors over three faces, I call it the CHRISTMAN X pattern. Inverting PA5 results in PA6. Also, PA7 has four colors over three faces, I call it the CHRISTMAN H pattern. See pattern P26, only the sides have an H pattern, the top and bottom have an X pattern. Inverting PA7 results in a six color arrangment with the top and bottom still being an H pattern, see PA8.

The main reason for me to take the CUBE apart to make the PA1 pattern was, as indicated in the CUBIK story, the result of spending many hours trying to reverse two edge cubes which would have made the CHRISTMAN DOTS pattern, but it just cannot be done (unless the CUBE is altered).

Pattern PA9 is another cinfiguration made from the altered CUBE. Since it has four colors over three faces, I call it the CHRISTMAN DOT CROSS pattern. Inverting PA9 results in pattern PA10.

```
PA2      .-----------,
        / b / t / b /|
       /---/---/---/ f:
      / t / t / t /|/|
     /---/---/---/ k/ k:
    / b / t / b /|/|/|
   :-----------: f/ r/ f:
   | r | e | r |/|/|/
   :-----------: k/ k/
   | e | f | e |/|/
   :-----------: f/
   | r | e | r |/ PA1
   '-----------' Inverted

PA4      .-----------,
        / b / b / b /|
       /---/---/---/ f:
      / b / f / b /|/|
     /---/---/---/ f/ f:
    / b / b / b /|/|/|
   :-----------: f/ t/ f:
   | r | r | r |/|/|/
   :-----------: f/ f/
   | r | r | r |/|/
   :-----------: f/
   | r | r | r |/ PA3 (DOT
   '-----------' Sequence-2)

PA6      .-----------,
        / t / t / t /|
       /---/---/---/ r:
      / t / t / t /|/|
     /---/---/---/ k/ k:
    / t / t / t /|/|/|
   :-----------: r/ r/ r:
   | f | e | f |/|/|/
   :-----------: k/ k/
   | e | f | e |/|/
   :-----------: r/
   | f | e | f |/ PA5
   '-----------' Inverted

PA8      .-----------,
        / t / b / t /|
       /---/---/---/ r:
      / t / t / t /|/|
     /---/---/---/ e/ k:
    / t / b / t /|/|/|
   :-----------: r/ r/ r:
   | f | k | f |/|/|/
   :-----------: k/ e/
   | e | f | e |/|/
   :-----------: r/
   | f | k | f |/ PA7
   '-----------' Inverted
```

```
PA9        .-----------.                     PA10       .-----------.
         / b / t / b / |                              / b / b / b / |
        /---/---/---/ f :                            /---/---/---/ f :
       / t / t / t /|/ |                            / b / t / b /|/ |
      /---/---/---/ f/ f :                          /---/---/---/ k / k :
     / b / t / b /|/|/ |                           / b / b / b /|/|/ |
    :-----------: f/ r/ f :                        :-----------: f/ r/ f :
    | r | r | r|/|/|/                              | r | e | r |/|/|/
    :-----------: f/ f/                            :-----------: k/ k/
    | r | f | r|/|/                                | e | f | e |/|/
    :-----------: f/                               :-----------: f/
    | r | r | r | r|/  CHRISTMAN                    | r | e | r | |/   PA9
    '-----------'      DOT CROSS                    '-----------'      Inverted
```

= =

N O T E S

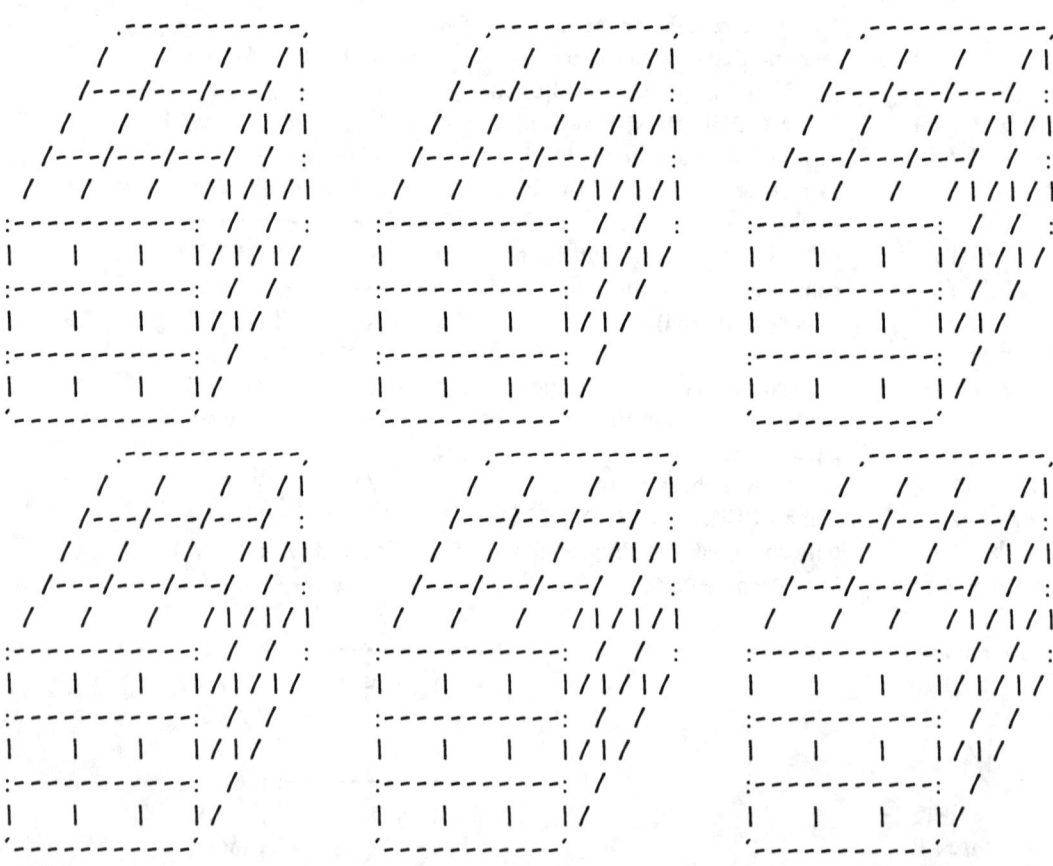

```
           .----------.
         /   /   /   /\
        /---/---/---/  :
       /   /   /   /\/\
      /---/---/---/  /  :
     /   /   /   /\/\/\
    :----------:  /  /  :
    |   |   |   |/\/\/\/
    :----------:  /  /
    |   |   |   |  /\/\/
    :----------:  /
    |   |   |   |  /
    '----------'
```

```
           .----------.
         /   /   /   /\
        /---/---/---/  :
       /   /   /   /\/\
      /---/---/---/  /  :
     /   /   /   /\/\/\
    :----------:  /  /  :
    |   |   |   |/\/\/\/
    :----------:  /  /
    |   |   |   |  /\/\/
    :----------:  /
    |   |   |   |  /
    '----------'
```

```
           .----------.
         /   /   /   /\
        /---/---/---/  :
       /   /   /   /\/\
      /---/---/---/  /  :
     /   /   /   /\/\/\
    :----------:  /  /  :
    |   |   |   |/\/\/\/
    :----------:  /  /
    |   |   |   |  /\/\/
    :----------:  /
    |   |   |   |  /
    '----------'
```

```
           .----------.
         /   /   /   /\
        /---/---/---/  :
       /   /   /   /\/\
      /---/---/---/  /  :
     /   /   /   /\/\/\
    :----------:  /  /  :
    |   |   |   |/\/\/\/
    :----------:  /  /
    |   |   |   |  /\/\/
    :----------:  /
    |   |   |   |  /
    '----------'
```

```
           .----------.
         /   /   /   /\
        /---/---/---/  :
       /   /   /   /\/\
      /---/---/---/  /  :
     /   /   /   /\/\/\
    :----------:  /  /  :
    |   |   |   |/\/\/\/
    :----------:  /  /
    |   |   |   |  /\/\/
    :----------:  /
    |   |   |   |  /
    '----------'
```

```
           .----------.
         /   /   /   /\
        /---/---/---/  :
       /   /   /   /\/\
      /---/---/---/  /  :
     /   /   /   /\/\/\
    :----------:  /  /  :
    |   |   |   |/\/\/\/
    :----------:  /  /
    |   |   |   |  /\/\/
    :----------:  /
    |   |   |   |  /
    '----------'
```

```
           .----------.
         /   /   /   /\
        /---/---/---/  :
       /   /   /   /\/\
      /---/---/---/  /  :
     /   /   /   /\/\/\
    :----------:  /  /  :
    |   |   |   |/\/\/\/
    :----------:  /  /
    |   |   |   |  /\/\/
    :----------:  /
    |   |   |   |  /
    '----------'
```

```
           .----------.
         /   /   /   /\
        /---/---/---/  :
       /   /   /   /\/\
      /---/---/---/  /  :
     /   /   /   /\/\/\
    :----------:  /  /  :
    |   |   |   |/\/\/\/
    :----------:  /  /
    |   |   |   |  /\/\/
    :----------:  /
    |   |   |   |  /
    '----------'
```

Notes

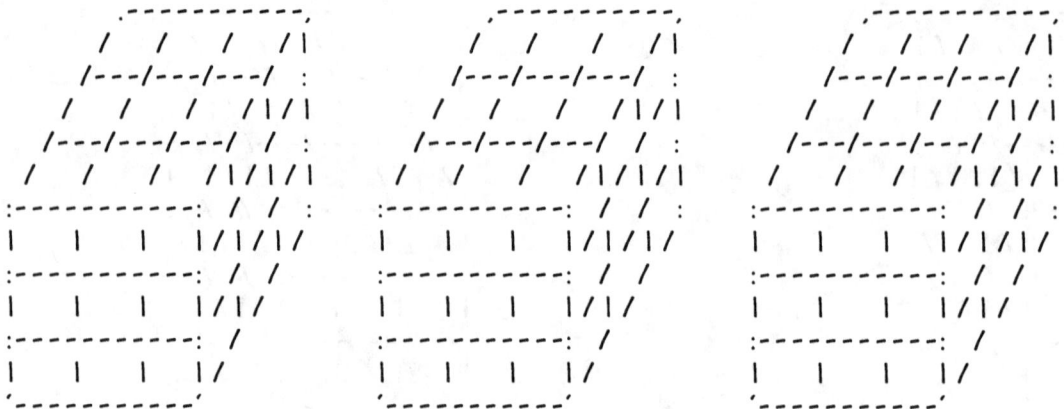

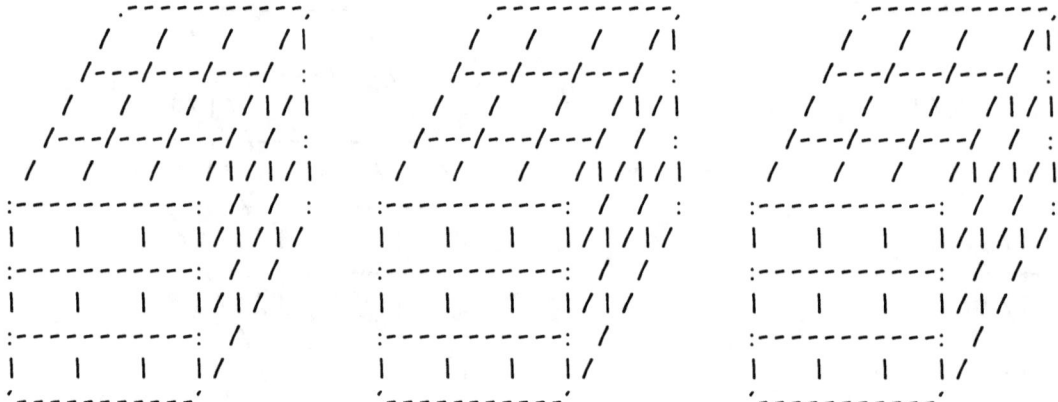

```
        .----------,
       /  /  /  /|
      /---/---/---/  :
     /  /  /  /  /|\|
    /---/---/---/  /  :
   /  /  /  /  /|\|/|
  :----------:  /  /  :
  |  |  |  |  |/|/|/|
  :----------:  /  /
  |  |  |  |  |/|/
  :----------:  /
  |  |  |  |  |/
  '----------'
```

```
            .----------,
           /  /  /  /|
          /---/---/---/  :
         /  /  /  /  /|\|
        /---/---/---/  /  :
       /  /  /  /  /|\|/|
      :----------:  /  /  :
      |  |  |  |  |/|/|/|
      :----------:  /  /
      |  |  |  |  |/|/
      :----------:  /
      |  |  |  |  |/
      '----------'
```

```
              .----------,
             /  /  /  /|
            /---/---/---/  :
           /  /  /  /  /|\|
          /---/---/---/  /  :
         /  /  /  /  /|\|/|
        :----------:  /  /  :
        |  |  |  |  |/|/|/|
        :----------:  /  /
        |  |  |  |  |/|/
        :----------:  /
        |  |  |  |  |/
        '----------'
```

```
          .-----------.
         /   /   /   /|
        /---/---/---/ :
       /   /   /   /|\|\
      /---/---/---/ / :
     /   /   /   /|\|\|\
    :-----------: / / :
    |   |   |   |/|/|/|/
    :-----------: / /
    |   |   |   |/|/
    :-----------: /
    |   |   |   |/
    '-----------'
          .-----------.
         /   /   /   /|
        /---/---/---/ :
       /   /   /   /|\|\
      /---/---/---/ / :
     /   /   /   /|\|\|\
    :-----------: / / :
    |   |   |   |/|/|/|/
    :-----------: / /
    |   |   |   |/|/
    :-----------: /
    |   |   |   |/
    '-----------'
          .-----------.
         /   /   /   /|
        /---/---/---/ :
       /   /   /   /|\|\
      /---/---/---/ / :
     /   /   /   /|\|\|\
    :-----------: / / :
    |   |   |   |/|/|/|/
    :-----------: / /
    |   |   |   |/|/
    :-----------: /
    |   |   |   |/
    '-----------'
          .-----------.
         /   /   /   /|
        /---/---/---/ :
       /   /   /   /|\|\
      /---/---/---/ / :
     /   /   /   /|\|\|\
    :-----------: / / :
    |   |   |   |/|/|/|/
    :-----------: / /
    |   |   |   |/|/
    :-----------: /
    |   |   |   |/
    '-----------'
```

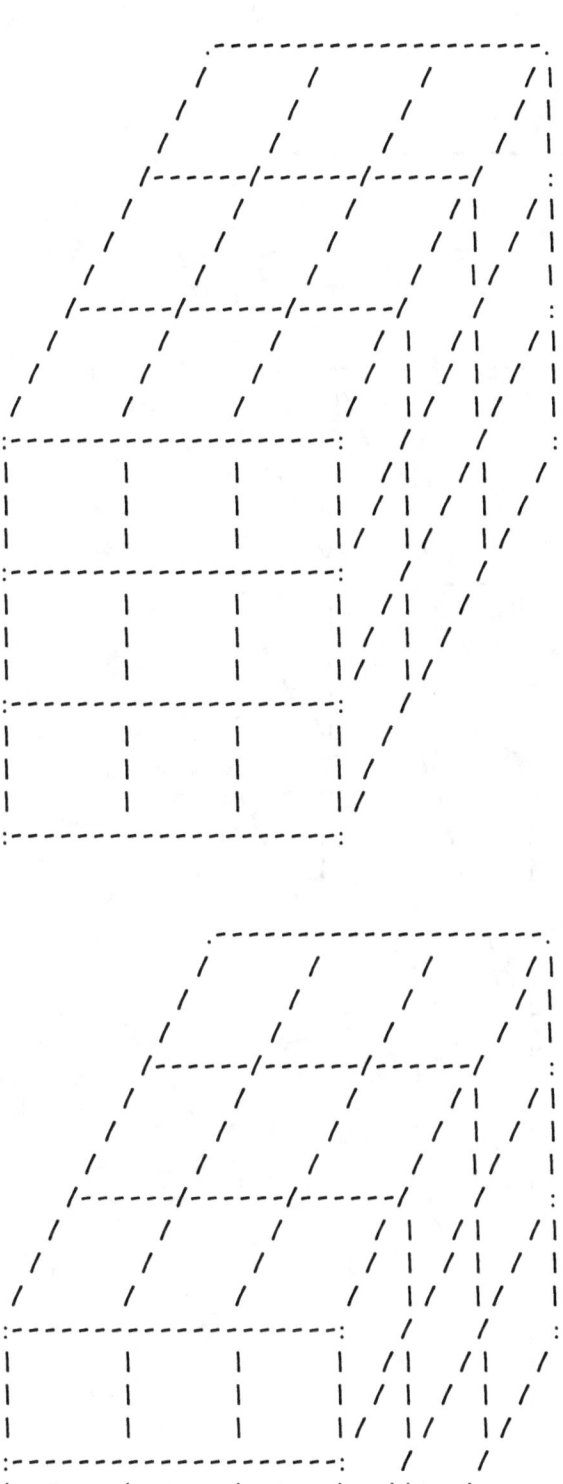

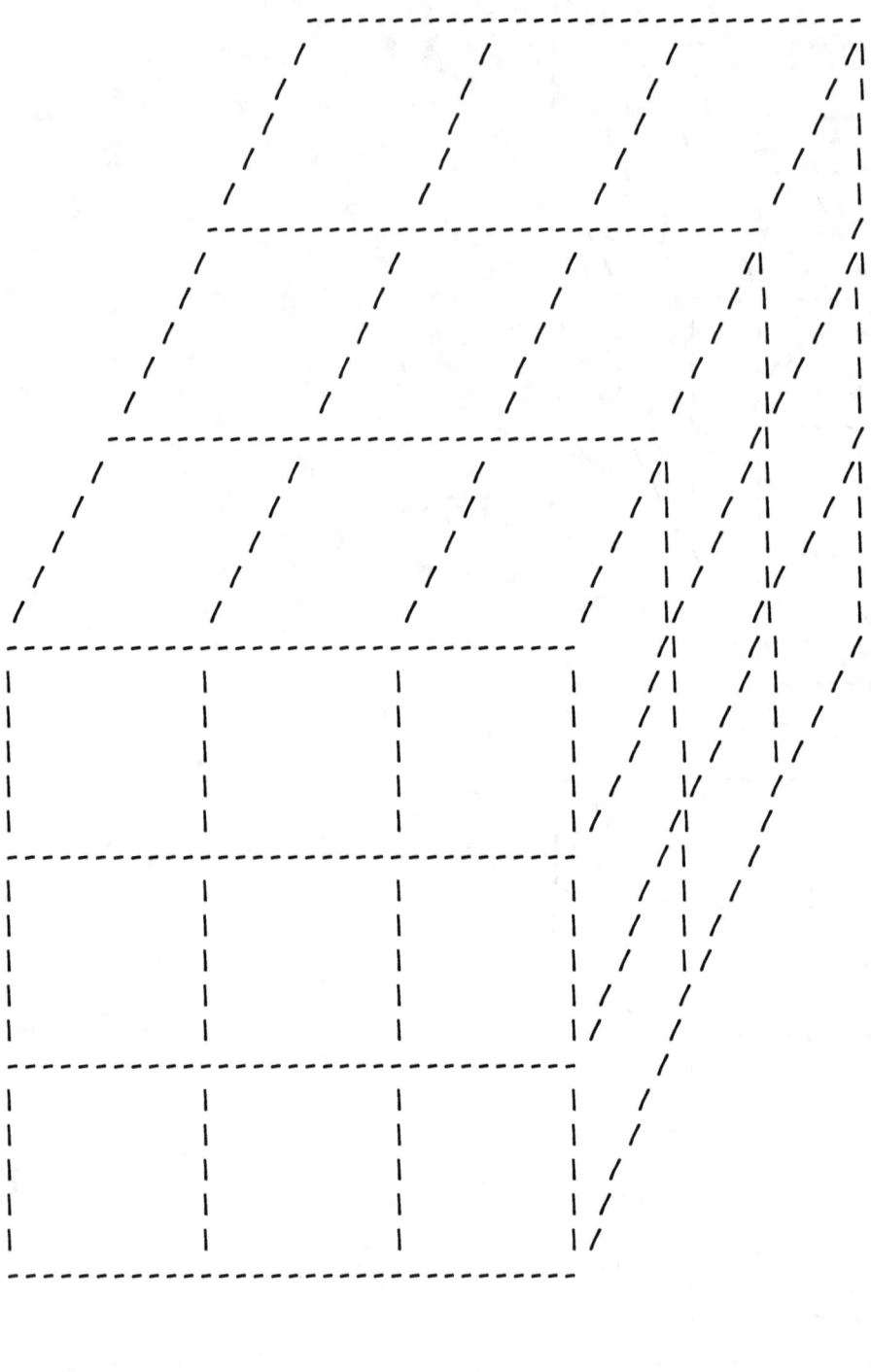

F A V O R I T E

```
TV        .-----------,
         / t / r / t /|
        /---/---/---/ r:
       / e / t / r /|/|
      /---/---/---/ k/ k:
     / t / e / t /|/|/|
     :-----------: r/ r/ r:
     | f | t | f |/|/|/
     :-----------: b/ b/
     | t | f | r |/|/
     :-----------: r/
     | f | r | f |/ four
     '-----------' corners

BV        .-----------,
         / b / f / b /|
        /---/---/---/ k:
       / f / b / k /|/|
      /---/---/---/ e/ t:
     / b / k / b /|/|/|
     :-----------: k/ k/ k:
     | e | b | e |/|/|/
     :-----------: e/ t/
     | f | e | b |/|/
     :-----------: k/
     | e | f | e |/ four
     '-----------' corners

TV        .-----------,
         / f / t / f /|
        /---/---/---/ t:
       / t / t / f /|/|
      /---/---/---/ t/ r:
     / f / f / f /|/|/|
     :-----------: t/ r/ t:
     | r | r | r |/|/|/
     :-----------: t/ r/
     | f | f | r |/|/
     :-----------: t/
     | r | f | r |/ worm
     '-----------' (phase 1)

BV        .-----------,
         / k / b / k /|
        /---/---/---/ e:
       / b / b / k /|/|
      /---/---/---/ e/ k:
     / k / k / k /|/|/|
     :-----------: e/ k/ e:
     | b | b | b |/|/|/
     :-----------: e/ k/
     | e | e | b |/|/
     :-----------: e/
     | b | e | b |/ worm
     '-----------' (phase 1)
```

2002 FAVORITES

Five favorite patterns all made from a scrambled CUBE are the 'ring-around-four-corners' pattern that was alluded to on page 29, a 'serpentine' pattern which seems to be a version of the 'ring-around-the-corner' pattern, and two patterns mentioned in Taylor/Rylands book and in the Scientific American article, namely, the 'snake' and the 'worm'. Two phases of the 'serpentine' pattern are shown in the Taylor/Rylands book as "large circles" of six edge cubes. Like the 'ring-around-the-corner' pattern the 'worm' pattern is cyclic in three repetitions and has three colors over three faces.

It was while generating the "four-corners" pattern from a scrambled CUBE that for the first time doing ALGO1 became confusing, hence I added the italics notes on page 27 to help.

```
TV        ,-----------,
         / r / t / r /|
        /---/---/---/ f:
       / t / t / r /|/|
      /---/---/---/ f/ r:
     / r / r / r /|/|/|
     :-----------: f/ r/ f:
     | t | t | t |/|/|/
     :-----------: f/ r/
     | f | f | t |/|/
     :-----------: f/
     | t | f | t |/ worm
     '-----------' (phase 2)

BV        ,-----------,
         / e / b / e /|
        /---/---/---/ b:
       / b / b / e /|/|
      /---/---/---/ b/ k:
     / e / e / e /|/|/|
     :-----------: b/ k/ b:
     | k | k | k |/|/|/
     :-----------: b/ k/
     | e | e | k |/|/
     :-----------: b/
     | k | e | k |/ worm
     '-----------' (phase 2)
```

```
TV        ,-----------,
         / t / t / t /|
        /---/---/---/ r:
       / t / t / f /|/|
      /---/---/---/ e/ e:
     / t / f / t /|/|/|
     :-----------: r/ r/ r:
     | f | t | f |/|/|/
     :-----------: r/ r/
     | t | f | f |/|/
     :-----------: r/
     | f | f | f |/ 'ser-
     '-----------' pentine'

BV        ,-----------,
         / b / b / b /|
        /---/---/---/ k:
       / b / b / k /|/|
      /---/---/---/ b/ b:
     / b / k / b /|/|/|
     :-----------: k/ k/ k:
     | e | r | e |/|/|/
     :-----------: k/ k/
     | r | e | e |/|/
     :-----------: k/
     | e | e | e |/ 'ser-
     '-----------' pentine'

TV        ,-----------,
         / b / b / b /|
        /---/---/---/ k:
       / t / t / t /|/|
      /---/---/---/ r/ k:
     / b / b / b /|/|/|
     :-----------: k/ r/ k:
     | e | e | e |/|/|/
     :-----------: r/ k/
     | e | f | f |/|/
     :-----------: k/
     | e | f | e |/ snake
     '-----------' snake

BV        ,-----------,
         / t / t / t /|
        /---/---/---/ r:
       / b / b / b /|/|
      /---/---/---/ k/ r:
     / t / t / t /|/|/|
     :-----------: r/ k/ r:
     | f | f | f |/|/|/
     :-----------: k/ r/
     | f | e | e |/|/
     :-----------: r/
     | f | e | f |/ snake
     '-----------' snake
```

ADDITIONAL PATTERNS (2002)

The following two pattern sections are a 2002 addition. The first pattern section shows the cyclic nature of my seven CUBIK algorithms that have been used to solve the CUBE, unscramble it. The second pattern section shows some exercises that solve the CUBE using the CUBIK solution. A table is included that tabulates the nineteen different flowchart paths germane to the CUBIK solution for these exercises. Only three of the ten exercises are at first diagrammed here in some detail since they basically just depict the essence of the CUBIK solution, that is, essentially solving the CUBE one cube at a time.

Of the seven CUBIK algorithms, four of them I call my translations and three of them I call my slice operations. All three slice operations (R/0, RR/0, RM/0) start with the right slice. The four translations are on reference cubes 1, 2, 3, and 4, namely, R/B/1/1, R/V/1/1, C/B/1/2, and C/V/1/2 respectively. As can be deduced, for solving the four bottom corner cubes, R/0 or RR/0 positions them where as R/B/1/1 orients them; for the four bottom edge cubes, R/V/1/1 both positions and orients them; and for the four middle edge cubes, C/V/1/2 or RM/0 positions them where as ALGO1 orients them. (ALGO1 is composed of C/B/1/2 and R/V/1/1 translations.) As can be seen these cyclic pattern figures show all six faces of the CUBE.

Patterns PC1, PC2, PC3, and PC4 show that translations on reference cube "1" (the asterisk cube) called translation R/B/1/1 but here abbreviated as "RB" is cyclic in three cycles. PC1 is the pristine CUBE, PC2 is the CUBE after one R/B/1/1 (8 step) translation, PC3 is after the second R/B/1/1 translation, and PC4 is after the third R/B/1/1 translation that returns the CUBE to a pristine state. Patterns PC2 and PC3 show that this translation (R/B/1/1) only churns cubes in the bottom layer, however, bottom corner cube f, e, b is arbitrarily my reference cube "0" (shown in parenthesis) and remains 'fixed' after the translation. Also by happenstance bottom edge cube e, b remains fixed so that only six of the eight-bottom layer cubes get churned (repositioned and/or reoriented).

Patterns PC4 through PC9 show that translations on reference cube "2" (the asterisk cube) called translation R/V/1/1 but here abbreviated "RV" is cyclic in five cycles. Patterns PC5, PC6, PC7, and PC8 show that this translation (R/V/1/1) churns the four middle edge cubes plus the right bottom edge cube. As before the bottom corner cube t, e, b remains 'fixed' and is shown in parenthesis. R/V/1/1 is a seven-step translation.

Patterns PC9 through PC12 show that translations on reference cube "3" (the asterisk cube) called translation C/B/1/2 but here abbreviated "CB" is cyclic in three cycles. Patterns PC10 and PC11 show that this translation (C/B/1/2) churns three of the bottom four edge cubes. The bottom edge cube f, b remains 'fixed' and is shown in parenthesis. C/B/1/2 is a seven-step translation.

Patterns PC12, PC13, and PC14 show that translations on reference cube "4" (the asterisk cube) called translation C/V/1/2 but here abbreviated "CV" is cyclic in two cycles. Pattern PC13 shows that this translation (C/V/1/2) interchanges diagonally the four middle edge cubes. The front center cube 'f remains 'fixed' and is shown in parenthesis. C/V/1/2 is an eight-step translation.

Patterns PC14 through PC21 depict that a slice operation on the right side called operation R/0 but here abbreviated "R" can be either an eight step or a nine-step operation. Patterns PC14 through PC17 show that the R/0-eight-stepper is cyclic in three cycles but is not preferred over the nine-stepper since the eight-stepper has no 'fixed' bottom front reference corner cube whereas the nine-stepper does. Therefore, the eight-stepper is shown for academic interests only.

(text continues on page PP6)

```
PC1    .-----------,              PC3    .-----------,              PC5    .-----------,
      / + / + / + /|                    / + / + / + /|                    / + / + / + /|
     /---/---/---/ r :                 /---/---/---/ r :                 /---/---/---/ r :
     / + / + / + /|/|                  / + / + / + /|/|                  / + / + / */|/|
    /---/---/---/ r / r :             /---/---/---/ r / r :             /---/---/---/ r / e :
   / + / + / * /|/|/|                / + / + / */|/|/|                 / + / + / + /|/|/|
  :-----------: r / r / r :         :-----------: r / r / b :         :-----------: r / r / r :
  | f | f | f |/|/|/                | f | f | f |/|/|/                | f | f | f |/|/|/
  :-----------: r / r /            :-----------: r / f /             :-----------: k / f /
  | f | f | f |/|/                 | f | f | f |/|/                  | e | f | r |/|/
  :-----------: r /                :-----------: f /                 :-----------: r /
  | (f) | f | f |/ tv              | (f) | k | b |/ tv               | (f) | f | f |/ tv
  '-----------'  RB(0)            '-----------'  RB(2)              '-----------'  RV(1)

PC1    .-----------,              PC3    .-----------,              PC5    .-----------,
      / b / b / b /|                    / r / b / k /|                    / b / r / b /|
     /---/---/---/ k :                 /---/---/---/ r :                 /---/---/---/ k :
     / b / b / b /|/|                  / b / b / b /|/|                  / b / b / b /|/|
    /---/---/---/ k / k :             /---/---/---/ r / k :             /---/---/---/ k / f :
   / b / b / b /|/|/|                / b / b / e /|/|/|                / b / b / b /|/|/|
  :-----------: k / k / k :         :-----------: b / k / k :         :-----------: k / k / k :
  | e | e | e |/|/|/                | e | e | k |/|/|/                | e | e | e |/|/|/
  :-----------: k / k /            :-----------: k / k /             :-----------: r / k /
  | e | e | e |/|/                 | e | e | e |/|/                  | k | e | b |/|/
  :-----------: k /                :-----------: k /                 :-----------: k /
  | e | e | e |/ bv               | e | e | e |/ bv                 | e | e | e |/ bv
  '-----------'  RB(0)            '-----------'  RB(2)              '-----------'  RV(1)

PC2    .-----------,              PC4    .-----------,              PC6    .-----------,
      / + / + / + /|                    / + / + / + /|                    / + / + / + /|
     /---/---/---/ r :                 /---/---/---/ r :                 /---/---/---/ r :
     / + / + / + /|/|                  / + / + / */|/|                  / + / + / */|/|
    /---/---/---/ r / r :             /---/---/---/ r / r :             /---/---/---/ r / k :
   / + / + / * /|/|/|                / + / + / + /|/|/|                / + / + / + /|/|/|
  :-----------: r / r / k :         :-----------: r / r / r :         :-----------: r / r / r :
  | f | f | f |/|/|/                | f | f | f |/|/|/                | f | f | f |/|/|/
  :-----------: r / k /            :-----------: r / r /             :-----------: f / r /
  | f | f | f |/|/                 | f | f | f |/|/                  | b | f | e |/|/
  :-----------: b /                :-----------: r / tv             :-----------: r /
  | (f) | r | r |/ tv              | (f) | f | f |/ RB(3)            | (f) | f | f |/ tv
  '-----------'  RB(1)            '-----------' or RV(0)            '-----------'  RV(2)

PC2    .-----------,              PC4    .-----------,              PC6    .-----------,
      / b / b / r /|                    / b / b / b /|                    / b / k / b /|
     /---/---/---/ b :                 /---/---/---/ k :                 /---/---/---/ k :
     / b / b / b /|/|                  / b / b / b /|/|                  / b / b / b /|/|
    /---/---/---/ f / k :             /---/---/---/ k / k :             /---/---/---/ k / e :
   / b / b / k /|/|/|                / b / b / b /|/|/|                / b / b / b /|/|/|
  :-----------: e / k / k :         :-----------: k / k / k :         :-----------: k / k / k :
  | e | e | b |/|/|/                | e | e | e |/|/|/                | e | e | e |/|/|/
  :-----------: k / k /            :-----------: k / k /             :-----------: f / k /
  | e | e | e |/|/                 | e | e | e |/|/                  | r | e | r |/|/
  :-----------: k /                :-----------: k / bv             :-----------: k /
  | e | e | e |/ bv               | e | e | e |/ RB(3)              | e | e | e |/ bv
  '-----------'  RB(1)            '-----------' or RV(0)            '-----------'  RV(2)
```

```
PC7      .-----------,        PC9      .-----------,        PC11     .-----------,
        / t / t / t /|               / t / t / t /|               / t / t / t /|
       /---/---/---/ r:             /---/---/---/ r:             /---/---/---/ r:
       / t / t / * /|/|             / t / t / t /|/|             / t / t / t /|/|
      /---/---/---/ r/ r:          /---/---/---/ r/ r:          /---/---/---/ r/ k:
      / t / t / t /|/|/|           / t / * / t /|/|/|           / t / * / t /|/|/|
     :-----------: r/ r/ r:       :-----------: r/ r/ r:       :-----------: r/ r/ r:
     | f | f | f |/|/|/           | f | f | f |/|/|/           | f | f | f |/|/|/
     :-----------: e / e /        :-----------: r / r /        :-----------: f / e /
     | r | f | k |/|/             | f | f | f |/|/             | f | f | f |/|/
     :-----------: r /            :-----------: r/ tv          :-----------: r /
     | (f) | f | f |/  tv         | f | (f) | f |/  RV(5)      | f | (f) | f |/  tv
     '-----------'  RV(3)         '-----------' or CB(0)       '-----------'  CB(2)

PC7      .-----------,        PC9      .-----------,        PC11     .-----------,
        / b / f / b /|               / b / b / b /|               / b / b / b /|
       /---/---/---/ k:             /---/---/---/ k:             /---/---/---/ k:
       / b / b / b /|/|             / b / b / b /|/|             / b / b / r /|/|
      /---/---/---/ k/ b:          /---/---/---/ k/ k:          /---/---/---/ b/ k:
      / b / b / b /|/|/|           / b / b / b /|/|/|           / b / k / b /|/|/|
     :-----------: k/ k/ k:       :-----------: k/ k/ k:       :-----------: k/ k/ k:
     | e | e | e |/|/|/           | e | e | e |/|/|/           | e | b | e |/|/|/
     :-----------: r / k /        :-----------: k / k /        :-----------: k / k /
     | f | e | k |/|/             | e | e | e |/|/             | e | e | e |/|/
     :-----------: k /            :-----------: k/ bv          :-----------: k /
     | e | e | e |/ bv           | e | e | e |/ RV(5)         | e | e | e |/ bv
     '-----------'  RV(3)         '-----------' or CB(0)       '-----------'  CB(2)

PC8      .-----------,        PC10     .-----------,        PC12     .-----------,
        / t / t / t /|               / t / t / t /|               / t / t / t /|
       /---/---/---/ r:             /---/---/---/ r:             /---/---/---/ r:
       / t / t / * /|/|             / t / t / t /|/|             / t / * / t /|/|
      /---/---/---/ r/ f:          /---/---/---/ r/ r:          /---/---/---/ r/ r:
      / t / t / t /|/|/|           / t / * / t /|/|/|           / t / t / t /|/|/|
     :-----------: r/ r/ r:       :-----------: r/ r/ r:       :-----------: r/ r/ r:
     | f | f | f |/|/|/           | f | f | f |/|/|/           | f | f | f |/|/|/
     :-----------: b / k /        :-----------: r / b /        :-----------: r / r /
     | k | f | r |/|/             | f | f | f |/|/             | f | (f) | f |/|/
     :-----------: r /            :-----------: r /            :-----------: r/ tv
     | (f) | f | f |/  tv         | f | (f) | f |/  tv         | f | f | f |/ CB(3)
     '-----------'  RV(4)         '-----------'  CB(1)         '-----------' or CV(0)

PC8      .-----------,        PC10     .-----------,        PC12     .-----------,
        / b / e / b /|               / b / k / b /|               / b / b / b /|
       /---/---/---/ k:             /---/---/---/ k:             /---/---/---/ k:
       / b / b / b /|/|             / b / b / e /|/|             / b / b / b /|/|
      /---/---/---/ k/ r:          /---/---/---/ b/ k:          /---/---/---/ k/ k:
      / b / b / b /|/|/|           / b / b / b /|/|/|           / b / b / b /|/|/|
     :-----------: k/ k/ k:       :-----------: k/ k/ k:       :-----------: k/ k/ k:
     | e | e | e |/|/|/           | e | r | e |/|/|/           | e | e | e |/|/|/
     :-----------: e / k /        :-----------: k / k /        :-----------: k / k /
     | r | e | f |/|/             | e | e | e |/|/             | e | e | e |/|/
     :-----------: k /            :-----------: k /            :-----------: k/ bv
     | e | e | e |/ bv           | e | e | e |/ bv           | e | e | e |/ CB(3)
     '-----------'  RV(4)         '-----------'  CB(1)         '-----------' or CV(0)
```

```
PC13      .-----------,        PC15      .-----------,        PC17      ,-----------,
         / t / t / t /|                 / t / t / * /|                 / t / t / * /|
        /---/---/---/ r :               /---/---/---/ r :               /---/---/---/ r :
       / t / * / t /|/|               / t / t / * /|/|               / t / t / * /|/|
      /---/---/---/ r / e :           /---/---/---/ r / r :           /---/---/---/ r / r :
     / t / t / t /|/|/|             / t / t / * /|/|/|             / t / t / * /|/|/|
    :-----------: r / r / r :       :-----------: r / r / r :       :-----------: r / r / r :
    | f | f | f |/|/|/|             | f | f | f |/|/|/|             | f | f | f |/|/|/|
    :-----------: e / r /           :-----------: r / b /           :-----------: r / r /
    | k | (f) | k |/|/|             | f | f | f |/|/|               | f | f | f |/|/|
    :-----------: r /               :-----------: b /               :-----------: r / tv
    | f | f | f |/ tv               | f | f | e |/ tv               | (f) | f | f | f |/ R(3)8
    '-----------'  CV(1)            '-----------'  R(1)8            '-----------' or R(0)9

PC13      .-----------,            PC15    ,-----------,          PC17      .-----------,
         / b / b / b /|                 / k / k / b /|                 / b / b / b /|
        /---/---/---/ k :               /---/---/---/ k :               /---/---/---/ k :
       / b / b / b /|/|               / b / b / b /|/|               / b / b / b /|/|
      /---/---/---/ k / f :           /---/---/---/ e / k :           /---/---/---/ k / k :
     / b / b / b /|/|/|             / r / r / b /|/|/|             / b / b / b /|/|/|
    :-----------: k / k / k :       :-----------: e / k / k :       :-----------: k / k / k :
    | e | e | e |/|/|/|             | b | b | f |/|/|/|             | e | e | e |/|/|/|
    :-----------: f / k /           :-----------: k / k /           :-----------: k / k /
    | r | e | r |/|/|               | e | e | e |/|/|               | e | e | e |/|/|
    :-----------: k /               :-----------: k /               :-----------: k / bv
    | e | e | e |/ bv               | e | e | e |/ bv               | e | e | e |/ R(3)8
    '-----------'  CV(1)            '-----------'  R(1)8            '-----------' or R(0)9

PC14      .-----------,          PC16    ,-----------,          PC18      ,-----------,
         / t / t / * /|                 / t / t / * /|                 / t / t / * /|
        /---/---/---/ r :               /---/---/---/ r :               /---/---/---/ r :
       / t / t / * /|/|               / t / t / * /|/|               / t / t / * /|/|
      /---/---/---/ r / r :           /---/---/---/ r / r :           /---/---/---/ r / r :
     / t / t / * /|/|/|             / t / t / * /|/|/|             / t / t / * /|/|/|
    :-----------: r / r / r :       :-----------: r / r / r :       :-----------: r / r / e :
    | f | f | f |/|/|/|             | f | f | f |/|/|/|             | f | f | f |/|/|/|
    :-----------: r / r /           :-----------: r / b /           :-----------: r / f /
    | f | f | f |/|/|               | f | f | f |/|/|               | f | f | f |/|/|
    :-----------: r / tv            :-----------: b /               :-----------: f /
    | f | f | f |/ CV(2)            | e | f | f |/ tv               | (f) | b | b |/ tv
    '-----------' or R(0)8          '-----------'  R(2)8            '-----------'  R(1)9

PC14      .-----------,          PC16    ,-----------,          PC18      ,-----------,
         / b / b / b /|                 / e / e / b /|                 / r / b / k /|
        /---/---/---/ k :               /---/---/---/ k :               /---/---/---/ b :
       / b / b / b /|/|               / b / b / r /|/|               / r / b / k /|/|
      /---/---/---/ k / k :           /---/---/---/ b / k :           /---/---/---/ b / k :
     / b / b / b /|/|/|             / b / b / r /|/|/|             / b / b / b /|/|/|
    :-----------: k / k / k :       :-----------: b / k / k :       :-----------: r / k / k :
    | e | e | e |/|/|/|             | k | k | f |/|/|/|             | e | e | k |/|/|/|
    :-----------: k / k /           :-----------: k / k /           :-----------: k / k /
    | e | e | e |/|/|               | e | e | e |/|/|               | e | e | e |/|/|
    :-----------: k / bv            :-----------: k /               :-----------: k /
    | e | e | e |/ CV(2)            | e | e | e |/ bv               | e | e | e |/ bv
    '-----------' or R(0)8          '-----------'  R(2)8            '-----------'  R(1)9
```

```
PC19      .-----------,          PC21      .-----------,          PC23      .-----------,
         / † / † / * /|                   / † / † / * /|                   / † / † / * /|
        /---/---/---/ r :                 /---/---/---/ r :                 /---/---/---/ r :
        / † / † / * /|/|                  / † / † / * /|/|                  / † / † / * /|/|
       /---/---/---/ r / r :              /---/---/---/ r / r :             /---/---/---/ r / r :
      / † / † / * /|/|/|                  / † / † / * /|/|/|                / † / † / * /|/|/|
     :-----------: r / r / r :           :-----------: r / r / r :        :-----------: r / r / r :
     | f | f | f | f |/|/|/              | f | f | f | f |/|/|/           | f | f | f | f |/|/|/
     :-----------: r / b /              :-----------: r / r /           :-----------: r / b /
     | f | f | f | f |/|/                | f | f | f | f |/|/             | f | f | f | f |/|/
     :-----------: r /                  :-----------: r /  tv           :-----------: r /
     | (f) | b | f |/  tv               | f | f | f | f |/   R(12)q      | f | f | (f) |/  tv
     '-----------'  R(6)q               '-----------'  or RR(0)         '-----------'  RR(6)


PC19      .-----------,          PC21      ,-----------,          PC23      ,-----------,
         / b / r / b /|                   / b / b / b /|                   / b / r / b /|
        /---/---/---/ k :                 /---/---/---/ k :                 /---/---/---/ k :
        / f / b / b /|/|                  / b / b / b /|/|                  / b / b / b /|/|
       /---/---/---/ k / k :              /---/---/---/ k / k :             /---/---/---/ k / k :
      / b / b / b /|/|/|                  / b / b / b /|/|/|                / b / e / b /|/|/|
     :-----------: k / k / k :           :-----------: k / k / k :        :-----------: k / k / k :
     | e | e | e | e |/|/|/              | e | e | e | e |/|/|/           | e | b | e | e |/|/|/
     :-----------: k / k /              :-----------: k / k /           :-----------: k / k /
     | e | e | e | e |/|/                | e | e | e | e |/|/             | e | e | e | e |/|/
     :-----------: k /                  :-----------: k /  bv           :-----------: k /
     | e | e | e | e |/  bv             | e | e | e | e |/   R(12)q      | e | e | e | e |/  bv
     '-----------'  R(6)q               '-----------'  or RR(0)         '-----------'  RR(6)


PC20      .-----------,          PC22      ,-----------,          PC24      ,-----------,
         / † / † / * /|                   / † / † / * /|                   / † / † / * /|
        /---/---/---/ r :                 /---/---/---/ r :                 /---/---/---/ r :
        / † / † / * /|/|                  / † / † / * /|/|                  / † / † / * /|/|
       /---/---/---/ r / r :              /---/---/---/ r / f :             /---/---/---/ r / k :
      / † / † / * /|/|/|                  / † / † / * /|/|/|                / † / † / * /|/|/|
     :-----------: r / r / k :           :-----------: r / r / f :        :-----------: r / r / e :
     | f | f | f | f |/|/|/              | f | f | f | f |/|/|/           | f | f | f | f |/|/|/
     :-----------: r / b /              :-----------: k / b /           :-----------: k / e /
     | f | f | f | f |/|/                | f | f | f | b |/|/             | f | f | f | r |/|/
     :-----------: b /                  :-----------: r /               :-----------: r /
     | (f) | r | r |/  tv               | k | f | (f) |/  tv            | r | f | (f) |/  tv
     '-----------'  R(11)q              '-----------'  RR(1)            '-----------'  RR(11)


PC20      .-----------,          PC22      ,-----------,          PC24      ,-----------,
         / f / f / b /|                   / b / e / e /|                   / b / b / b /|
        /---/---/---/ e :                 /---/---/---/ b :                 /---/---/---/ f :
        / b / b / k /|/|                  / b / b / k /|/|                  / b / b / f /|/|
       /---/---/---/ b / k :              /---/---/---/ r / r :             /---/---/---/ r / b :
      / b / b / k /|/|/|                  / b / b / e /|/|/|                / k / r / k /|/|/|
     :-----------: b / k / k :           :-----------: b / k / k :        :-----------: e / k / k :
     | e | e | r | r |/|/|/              | r | r | k | k |/|/|/           | b | b | b | b |/|/|/
     :-----------: k / k /              :-----------: k / k /           :-----------: k / k /
     | e | e | e | e |/|/                | e | e | e | e |/|/             | e | e | e | e |/|/
     :-----------: k /                  :-----------: k /               :-----------: k /
     | e | e | e | e |/  bv             | e | e | e | e |/  bv          | e | e | e | e |/  bv
     '-----------'  R(11)q              '-----------'  RR(1)            '-----------'  RR(11)
```

```
PC25        .-----------,
          / t / t / * /|
         /---/---/---/ r :
        / t / t / * /|/|
       /---/---/---/ r / r :
      / t / t / * /|/|/|
     :-----------: r / r / r :
     | f | f | f |/|/|/
     :-----------: r / r /
     | (f) | f | f |/|/
     :-----------: r / tv
     | f | f | f |/ RR(12)
     '-----------' or RM(0)

PC25        .-----------,
          / b / b / b /|
         /---/---/---/ k :
        / b / b / b /|/|
       /---/---/---/ k / k :
      / b / b / b /|/|/|
     :-----------: k / k / k :
     | e | e | e |/|/|/
     :-----------: k / k /
     | e | e | e |/|/
     :-----------: k / bv
     | e | e | e |/ RR(12)
     '-----------' or RM(0)

PC26        .-----------,
          / t / t / * /|
         /---/---/---/ r :
        / t / t / * /|/|
       /---/---/---/ r / e :
      / t / t / * /|/|/|
     :-----------: r / r / r :
     | f | f | f |/|/|/
     :-----------: r / r /
     | (f) | f | k |/|/
     :-----------: r /
     | f | f | f |/ tv
     '-----------' RM(1)

PC26        .-----------,
          / b / b / b /|
         /---/---/---/ k :
        / b / b / b /|/|
       /---/---/---/ k / k :
      / b / b / b /|/|/|
     :-----------: k / k / k :
     | e | e | e |/|/|/
     :-----------: f / k /
     | e | e | r |/|/
     :-----------: k /
     | e | e | e |/ bv
     '-----------' RM(1)
```

```
PC27        .-----------,
          / t / t / * /|
         /---/---/---/ r :
        / t / t / * /|/|
       /---/---/---/ r / r :
      / t / t / * /|/|/|
     :-----------: r / r / r :
     | f | f | f |/|/|/
     :-----------: e / r /
     | (f) | f | k |/|/
     :-----------: r /
     | f | f | f |/ tv
     '-----------' RM(2)

PC27        .-----------,
          / b / b / b /|
         /---/---/---/ k :
        / b / b / b /|/|
       /---/---/---/ k / f :
      / b / b / b /|/|/|
     :-----------: k / k / k :
     | e | e | e |/|/|/
     :-----------: k / k /
     | e | e | r |/|/
     :-----------: k /
     | e | e | e |/ bv
     '-----------' RM(2)

PC28        .-----------,
          / t / t / * /|
         /---/---/---/ r :
        / t / t / * /|/|
       /---/---/---/ r / r :
      / t / t / * /|/|/|
     :-----------: r / r / r :
     | f | f | f |/|/|/
     :-----------: r / r /
     | (f) | f | f |/|/
     :-----------: r /
     | f | f | f |/ tv
     '-----------' RM(3)

PC28        .-----------,
          / b / b / b /|
         /---/---/---/ k :
        / b / b / b /|/|
       /---/---/---/ k / k :
      / b / b / b /|/|/|
     :-----------: k / k / k :
     | e | e | e |/|/|/
     :-----------: k / k /
     | e | e | e |/|/
     :-----------: k /
     | e | e | e |/ bv
     '-----------' RM(3)
```

Patterns PC15 and PC16 show that an R/0 operation churns six of the eight bottom layer cubes. By happenstance the bottom edge cube f, b and corner cube r, k, b remains fixed. Patterns PC17 through PC21 show that the R/0 9-stepper is cyclic in twelve cycles but unlike the 8-stepper it has a bottom corner 'fixed' which is shown in parenthesis (the f, e, b corner). Only the start, first, sixth, eleventh, and twelfth cycles are diagrammed.

Now six of the eight bottom layer cubes are churned and by happenstance the bottom edge e, b cube remains fixed.

Patterns PC21 through PC25 depict that a slice operation on the right side called operation RR/0 but here abbreviated "RR" is cyclic in twelve cycles. Only the start, first, sixth, eleventh, and twelfth cycles are diagrammed. The RR/0 operation churns six of the eight bottom layer cubes plus two of the right middle edge cubes. By happenstance edge cube f, b remains fixed. The bottom right corner cube f, r, b remains fixed. RR/0 is a ten-step operation.

Patterns PC25 through PC28 show that a slice operation on the right side called RM/0 but here abbreviated as "RM" is cyclic in three cycles. Patterns PC26 and PC27 show that a RM/0 operation churns (rotates) three of the four middle edge cubes. The front left middle edge cube f, e remains fixed and is shown in parenthesis. RM/0 is a four step operation.

The next pattern section illustrates some examples of how the CUBIK algorithms come into play. The FOUR STEP APPROACH finishes the top; the R/0 or RR/0 slice operation and/or R/B/1/1 translation finishes the bottom four corner cubes; the R/V/1/1 translation finishes the bottom four edge cubes; and finally the RM/0 slice operation and/or C/V/1/2, R/V/1/1, and/or C/B/1/2 translations finishes the four middle edge cubes.

```
S1(tv)    .------------,          S2(tv)    ,------------,
        / b / e / f /|                    / t / t / b /|
       /---/---/---/ r :                 /---/---/---/ k :
      / e / t / f /|/|                   / t / t / f /|/|
     /---/---/---/ b / k:               /---/---/---/ b / k:
    / K / r / r /|/|/|                  / t / r / r /|/|/|
   :------------: t / r / e:           :------------: t / r / f:
   | e | t | K |/|/|/|                 | f | t | K |/|/|/|
   :------------: e / K/                :------------: e / f/
   | e | f | f |/|/|                   | r | f | f |/|/|
   :------------: f /                  :------------: b /
   | f | K | t |/ TEN                  | f | e | r |/ side
   '------------' TWISTS               '------------' t's done

S1(bv)    .------------,          S2(bv)    ,------------,
        / e / t / t /|                    / f / t / e /|
       /---/---/---/ K :                 /---/---/---/ b :
      / r / b / r /|/|                   / b / b / K /|/|
     /---/---/---/ f / b:               /---/---/---/ e / r:
    / b / b / r /|/|/|                  / t / r / e /|/|/|
   :------------: b / K/ t:            :------------: b / K / r:
   | e | r | f |/|/|/|                 | r | f | K |/|/|/|
   :------------: f / b/                :------------: K / K /
   | t | e | t |/|/|                   | b | e | b |/|/|
   :------------: K /                  :------------: K /
   | b | K | r |/ TEN                  | e | e | e |/ side
   '------------' TWISTS               '------------' t's done

S3(tv)    .------------,          S4(tv)    ,------------,
        / t / t / t /|                    / t / t / t /|
       /---/---/---/ r :                 /---/---/---/ r :
      / t / t / t /|/|                   / t / t / t /|/|
     /---/---/---/ r / r:               /---/---/---/ r / e:
    / t / t / t /|/|/|                  / t / t / t /|/|/|
   :------------: r / r / f:           :------------: r / r / b:
   | f | f | f |/|/|/|                 | f | f | f |/|/|/|
   :------------: f / f/                :------------: b / f/
   | r | f | e |/|/|                   | r | f | e |/|/|
   :------------: b /                  :------------: r /
   | b | r | K |/ top                  | b | f | f |/ cor.
   '------------' done                 '------------' pos. done

S3(bv)    .------------,          S4(bv)    ,------------,
        / r / b / b /|             r,f,b-/ b / r / K /|
       /---/---/---/ r :-r,f,b          /---/---/---/ r :
      / f / b / K /|/|                   / b / b / K /|/|
     /---/---/---/ e / K:               /---/---/---/ r / f:
    / K / e / e /|/|/|                  / e / K / b /|/|/|
   :------------: f / K / K:           :------------: K / K / K:
   | e | b | b |/|/|/|                 | f | e | e |/|/|/|
   :------------: K / K/                :------------: K / K /
   | b | e | b |/|/|                   | b | e | b |/|/|
   :------------: K /                  :------------: K /
   | e | e | e |/ top                  | e | e | e |/ cor.
   '------------' done                 '------------' pos. done
```

EXAMPLE #1 LET'S DO IT

Figure S1 is a scrambled CUBE (after ten twists with eyes closed). Figure S2 is the CUBE after moving four cubes that were on the sides of either the middle or the bottom layers. At the start sometimes the middle and bottom layers are moved together (don't care). Table S shows that 16 steps moved the four cubes. The correction step aligned the faces for the S2 figure.

To move the other four cubes, 23 steps were required since the two corner cubes had to be moved to the sides first. The t,r cube got moved to the side and was already in the READY position. It took a total of 39 steps plus one alignment to finish the top as shown in Figure S3.

By moving the bottom layer 1Qt to the left, the r,f,b corner cube shown in Figure S3 gets correctly positioned and oriented as a bottom corner refer- ence cube as shown in Figure S4. Then it can be deduced that the bottom corner cube e,k,b (shown in bold type) shown in Figure S3 is positioned correctly but not oriented correctly and is diago- nally across from the r,f,b refer- ence cube. The other two bottom corner cubes e,f,b and r,k,b are out of positioned and need to be inter- changed using an RR/O slice operation on the front face since the reference cube needs to be at the bottom front right for an RR/O operation. Figure S4 is the CUBE after the RR/O operation and now all four bottom corners are positioned correctly.

A perusal of Figure S4 shows that the two corners r,f,b and e,k,b are okay but the other two corners are not oriented correctly, therefore, the bottom four corners are not in-sync. An R/B/1/1 trans- lation is needed on one of the faces in order to get them in-sync. Experience has shown that an

```
S5(tv)      .-----------,
         / t / t / t /|
        /---/---/---/ r :
       / t / t / t /|/|
      /---/---/---/ r / e :
     / t / t / t /|/|/|
    :-----------: r / r / r :-rkb
    | f | f | f |/|/|/
    :-----------: b / r /
    | r | f | e |/|/
    :-----------: b /
    | b | f | r |/
    '-----------'  in-sync

S5(bv)      .-----------,
         / f / k / b /|
        /---/---/---/ k :-r,k,b
       / r / b / b /|/|
      /---/---/---/ f / f :
     / e / k / k /|/|/|
    :-----------: e / k / k :
    | f | e | b |/|/|/
    :-----------: k / k /
    | b | e | b |/|/
    :-----------: k /
    | e | e | e |/
    '-----------'  in-sync

S7(tv)      .-----------,
         / t / t / t /|
        /---/---/---/ r :
       / t / t / t /|/|
      /---/---/---/ r / e :
     / t / t / t /|/|/|
    :-----------: r / r / r :
    | f | f | f |/|/|/
    :-----------: b / r /
    | r | f | e |/|/
    :-----------: r /
    | f | e | f |/ corners
    '-----------'  done

S7(bv)      .-----------,
         / b / k / b /|
        /---/---/---/ k :
       / k / b / r /|/|
      /---/---/---/ f / f :
     / b / b / b /|/|/|
    :-----------: k / k / k :
    | e | f | e |/|/|/
    :-----------: k / k /
    | b | e | b |/|/
    :-----------: k /
    | e | e | e |/ corners
    '-----------'  done
```

```
S6(tv)      ,-----------,
         / t / t / t /|
        /---/---/---/ r :
       / t / t / t /|/|
      /---/---/---/ r / e :
     / t / t / t /|/|/|
    :-----------: r / r / r :
    | f | f | f |/|/|/
    :-----------: b / r /
    | r | f | e |/|/
    :-----------: f /
    | e | f | b |/ still
    '-----------'  in-sync

S6(bv)      ,-----------,
         / r / k / b /|
        /---/---/---/ k :
       / b / b / k /|/|
      /---/---/---/ e / f :
     / f / r / e /|/|/|
    :-----------: b / k / k :
    | b | f | k |/|/|/
    :-----------: k / k /
    | b | e | b |/|/
    :-----------: k /
    | e | e | e |/ still
    '-----------'  in-sync

S8(tv)      ,-----------,
         / t / t / t /|
        /---/---/---/ r :
       / t / t / t /|/|
      /---/---/---/ r / e :
     / b / k / b /|/|/|
    :-----------: e / r / r :
    | f | e | f |/|/|/
    :-----------: b / r /-r,k
    | e | f | r |/|/
    :-----------: e /
    | f | f | f |/ r,b
    '-----------'  READY

S8(bv)      ,-----------,
         / t / k / b /|
        /---/---/---/ k :
       / t / b / r /|/|
      /---/---/---/ f / f :
     / t / b / b /|/|/|
    :-----------: k / k / k :
    | r | f | e |/|/|/
    :-----------: k / k /
    | b | e | b |/|/
    :-----------: k /
    | r | e | e |/ r,b
    '-----------'  READY
```

R/B/1/1 translation on the front face 'f' gets them in-sync. Figure S5 is the CUBE after this R/B/1/1 translation. Perusal of Figure S5 shows the r,k,b bottom corner to be okay and the other three bottom corners in-sync.

Now with r,k,b as a reference cube zero, two R/B/1/1 translations on the back 'k' face finishes the four bottom corner cubes where Figure S6 shows the CUBE after the first translation (three corners are still in-sync). Figure S7 is the CUBE with all eight corners done after the three R/B/1/1 translations.

A perusal of Figure S7 shows that all four bottom edge cubes are wrong. However, there are three bottom edge cubes already in the middle layer that can be sent down to the bottom layer with a R/V/1/1 translation but none are in a READY position. Figure S8 shows that by inverting the front 'f' face puts the r,b edge cube in a READY position which bumps back to the middle layer the r,k edge cube. Figure S9 shows the CUBE after R/V/1/1 translation on the front 'f' face (puts the r,b edge cube in the bottom layer) and the front face restored (inverted back). A perusal of Figure S9 shows that either inverting the right 'r' face (puts k,b edge cube in READY position) or the back 'k' face (puts e,b edge cube in READY position) bumps back either the r,f edge cube or the f,b edge cube to the middle layer. Let's bump back the r,f edge cube to the middle layer. Figure S10 shows the CUBE after a R/V/1/1 translation on an inverted right 'r' face (puts the k,b edge cube in the bottom layer) and after the right face is restored (inverted back). "TWO-MORE TO GO". This is amazing, a perusal of Figure S10 shows that edge cube e,b is already in the READY position (and so far I haven't needed to

```
S9(tv)    .-----------,
        / t / t / t /|
       /---/---/---/ r :
      / t / t / t /|/|
     /---/---/---/ r / b :
    / t / t / t /|/|/|
   :-----------: r / r / r :
   | f | f | f |/|/|/|
   :-----------: k / r /-r,b
   | e | f | b |/|/
   :-----------: r /
   | f | e | f |/ r,b
   '-----------' done

S9(bv)    .-----------,
        / b / b / b /|
       /---/---/---/ k :
      / k / b / r /|/|
     /---/---/---/ f / e :
    / b / b / b /|/|/|
   :-----------: k / k / k :
   | e | f | e |/|/|/|
   :-----------: r / k /
   | f | e | k |/|/
   :-----------: k /
   | e | e | e |/ r,b
   '-----------' done

S11(tv)   .-----------,
        / t / t / t /|
       /---/---/---/ r :
      / t / t / t /|/|
     /---/---/---/ r / k :
    / t / t / t /|/|/|
   :-----------: r / r / r :
   | f | f | f |/|/|/|
   :-----------: b / r /
   | f | f | f |/|/
   :-----------: r /
   | f | e | f |/ e,b
   '-----------' done

S11(bv)   .-----------,
        / b / b / b /|
       /---/---/---/ k :
      / k / b / b /|/|
     /---/---/---/ k / r :
    / b / b / b /|/|/|
   :-----------: k / k / k :
   | e | e | e |/|/|/|
   :-----------: f / k /
   | e | e | r |/|/
   :-----------: k /
   | e | e | e |/ e,b
   '-----------' done
```

```
S10(tv)   .-----------,
        / t / t / t /|
       /---/---/---/ r :
      / t / t / t /|/|
     /---/---/---/ r / e :
    / t / t / t /|/|/|
   :-----------: r / r / r :
   | f | f | f |/|/|/|
   :-----------: r / r /
   | r | f | k |/|/
   :-----------: r /
   | f | e | f |/ k,b
   '-----------' done

S10(bv)   .-----------,
        / b / b / b /|
       /---/---/---/ k :
      / k / b / b /|/|
     /---/---/---/ k / f :
    / b / b / b /|/|/|
   :-----------: k / k / k :
   | e | f | e |/|/|/|
   :-----------: e / k /
   | f | e | b |/|/
   :-----------: k /
   | e | e | e |/ k,b
   '-----------' done

S12(tv)   .-----------,
        / t / t / t /|
       /---/---/---/ r :
      / t / t / t /|/|
     /---/---/---/ r / f :
    / t / t / t /|/|/|
   :-----------: r / k / r :
   | f | f | f |/|/|/|
   :-----------: r / r /
f,b-| b | r | k |/|/
   :-----------: r /
   | f | e | f |/ f,b
   '-----------' READY

S12(bv)   .-----------,
        / b / b / b /|
       /---/---/---/ k :
  e,k-/ k / b / b /|/|
     /---/---/---/ k / r :
    / b / b / b /|/|/|
   :-----------: k / e / k :
   | e | e | e |/|/|/|
   :-----------: e / k /
   | f | f | f |/|/
   :-----------: k /
   | e | e | e |/ f,b
   '-----------' READY
```

adjust the middle layer, YET).
Figure S11 shows the CUBE after
a R/V/1/1 translation on the back
'k' face puts the e,b edge cube
in the bottom layer and bumps
back the f,b edge cube. Now the
the middle layer needs to be adjusted. A perusal of Figure S11 shows
that if the middle layer is adjusted 1Qt
to the left, edge cube f,b is in the READY
position. Figure S12 shows the f,b edge
cube in the READY position to
bump back the e,k edge cube to
the middle layer. Figure S13
shows the CUBE after a R/V/1/1
translation on the left face ('f
center cube now facing you)
bumps the e,k edge cube back to
the middle layer and the middle
is returned 1Qt to the right. The
bottom layer is now complete.
"FOUR MIDDLE EDGE CUBES TO
GO". A perusal of Figure S13 shows that
one of the four middle edge cubes is
already positioned correctly (but
not oriented), the e,k edge cube.

```
S13(tv)    .-----------,
         / t / t / t /|
        /---/---/---/ r :
       / t / t / t /|/|
      /---/---/---/ r / e :
     / t / t / t /|/|/|
    :-----------: r / r / r :
    | f | f | f |/|/|/|
    :-----------: r / r /
    | r | f | k |/|/
    :-----------: r /
    | f | f | f |/ bottom
    '-----------' done

S13(bv)    .-----------,
         / b / b / b /|
        /---/---/---/ k :
       / b / b / b /|/|
      /---/---/---/ k / f :
     / b / b / b /|/|/|
    :-----------: k / k / k :
    | e | e | e |/|/|/|
    :-----------: e / k /
    | f | e | k |/|/
    :-----------: k /
    | e | e | e |/ bottom
    '-----------' done
```

```
S14(tv)   .-----------,
        / t / t / t /|
       /---/---/---/ r :
      / t / t / t /|/|
     /---/---/---/ r / f :
    / t / t / t /|/|/|
   :-----------: r / r / r :
   | f | f | f |/|/|/|/
   :-----------: e / r /
   | K | f | f |/|/|/
   :-----------: r /
   | f | f | f |/ e,K ref.
   '-----------' 1st RM/o
S14(bv)   .-----------,
        / b / b / b /|
       /---/---/---/ K :
      / b / b / b /|/|
     /---/---/---/ K / r :
    / b / b / b /|/|/|
   :-----------: K / K / K :
   | e | e | e |/|/|/|/
   :-----------: e / K /
   | r | e | K |/|/|/
   :-----------: K /
   | e | e | e |/ e,K ref.
   '-----------' 1st RM/o
S16(tv)   .-----------,
        / b / t / t /|
       /---/---/---/ r :
      / b / t / t /|/|
     /---/---/---/ r / r :
    / b / t / t /|/|/|
   :-----------: r / r / r :
   | K | f | f |/|/|/|/
   :-----------: f / r /
   | e | f | r |/|/|/
   :-----------: r /
   | K | f | f |/ ALGO1
   '-----------' READY
S16(bv)   .-----------,
        / b / b / b /|
       /---/---/---/ K :
      / b / b / b /|/|
     /---/---/---/ K / K :
    / t / t / t /|/|/|
   :-----------: f / K / K :
   | e | e | e |/|/|/|/
   :-----------: f / K /
   | K | e | e |/|/|/
   :-----------: f /
   | e | e | e |/ ALGO1
   '-----------' READY
```

```
S15(tv)   ,-----------,
        / t / t / t /|
       /---/---/---/ r :
      / t / t / t /|/|
     /---/---/---/ r / r :
    / t / t / t /|/|/|
   :-----------: r / r / r :
   | f | f | f |/|/|/|/
   :-----------: f / r /
   | f | f | r |/|/|/
   :-----------: r /
   | f | f | f |/ middle
   '-----------' pos. done
S15(bv)   ,-----------,
        / b / b / b /|
       /---/---/---/ K :
      / b / b / b /|/|
     /---/---/---/ K / K :
    / b / b / b /|/|/|
   :-----------: K / K / K :
   | e | e | e |/|/|/|/
   :-----------: e / K /
   | e | e | K |/|/|/
   :-----------: K /
   | e | e | e |/ middle
   '-----------' pos. done
S17(tv)   ,-----------,
        / b / t / t /|
       /---/---/---/ r :
      / b / t / t /|/|
     /---/---/---/ r / r :
    / b / t / t /|/|/|
   :-----------: r / r / r :
   | K | f | f |/|/|/|/
   :-----------: e / r /-r,f
   | b | f | f |/|/|/
   :-----------: r /
   | K | f | f |/ e,K & r,f
   '-----------' on bottom
S17(bv)   ,-----------,
        / b / f / b /|
       /---/---/---/ K :
      / b / b / e /|/|
     /---/---/---/ K / b :
    / t / t / t /|/|/|
   :-----------: f / K / K :
   | e | e | e |/|/|/|/
   :-----------: r / K /
   | K | e | K |/|/|/
   :-----------: f /
   | e | e | e |/ e,K & r,f
   '-----------' on bottom
```

(Therefore, we don't have to do a C/V/1/2 translation.) We need a RM/o operation that rotates the other three middle edge cubes. Figure S14 shows the CUBE after a RM/o operation on the left 'e' face with edge cube e,K as the reference cube which remains fixed while the other three rotate. Figure S15 shows the CUBE after a second RM/o operation on the left 'e' face (again) which correctly positions all four middle edge cubes (does not have to orient them) with edge cube e,K remaining fixed.

Figure S15 shows that there remains two middle edge cubes not oriented correctly but they are not adjacent so they are position-ed diagonally instead. An ALGO1 operation orients them but they need to be adjacent. (If they were adjacent, their READY position is depicted in Figure 14, page 28.) Figure S16 shows the e,K and r,f edge cubes in a READY position after inverting the left side (similar to Figure 15, page 28) so that an ALGO1 oper-ation can begin as follows.

A R/V/1/1 translation on the front 'f' face sends the r,f edge cube to the bottom and then a R/V/1/1 translation on the right 'r' face sends the e,K edge cube to the bottom layer. Figure S17 shows these two edge cubes (e,K and r,f) in the bottom layer before a C/B/1/2 translation inverts them. Now a C/B/1/2 translation on the left 'e' face inverts them as shown by Figure S18. Figure S19 shows that after the middle layer is ad-justed 1Qt to the left, the r,b edge cube in the middle layer is the one and only one in a READY position to bump an edge cube in the bot-tom layer (which happens to be the e,K edge cube) by a R/V/1/1 translation on the front 'f' face (right 'r' center cube towards you). Figure S20 shows the CUBE after the first of these three bumping operations.

```
S18(tv)  .-----------,          S19(tv)  .-----------,          S20(tv)  .-----------,
        / b / t / t /|                  / b / t / t /|                  / b / t / t /|
       /---/---/---/ r:                /---/---/---/ r:                /---/---/---/ r:
      / b / t / t /|/|                / b / t / t /|/|                / b / t / t /|/|
     /---/---/---/ r/ r:            /---/---/---/ r/ r:            /---/---/---/ r/ f:
    / b / t / t /|/|/|            / b / t / t /|/|/|            / b / t / t /|/|/|
   :-----------: r/ K/ r:        :-----------: r/ K/ r:        :-----------: r/ K/ r:
   | K | f | f |/|/|/|           | K | f | f |/|/|/|           | K | f | f |/|/|/|
   :-----------: e/ e/-e,K       :-----------: b/ e/-e,K       :-----------: K/ r/
   | b | f | f |/|/              | e | r | r |/|/              | b | r | r |/|/
   :-----------: r/              :-----------: r/              :-----------: r/
   | K | f | f |/ e,K & r,f      | K | f | f |/ r,b            | K | f | f |/ r,b
   '-----------' inverted        '-----------' READY           '-----------' done

S18(bv)  .-----------,          S19(bv)  .-----------,          S20(bv)  .-----------,
        / b / K / b /|                  / b / K / b /|                  / b / b / b /|
       /---/---/---/ K:                /---/---/---/ K:                /---/---/---/ K:
  r,f-/ r / b / b /|/|            / r / b / b /|/|            / r / b / b /|/|
     /---/---/---/ f/ b:          /---/---/---/ f/ K:          /---/---/---/ f/ e:
    / t / t / t /|/|/|            / t / t / t /|/|/|            / t / t / t /|/|/|
   :-----------: f/ K/ K:        :-----------: f/ e/ K:        :-----------: f/ e/ K:
   | e | e | e |/|/|/|           | e | e | e |/|/|/|           | e | e | e |/|/|/|
   :-----------: r/ K/            :-----------: K/ K/            :-----------: e/ K/
   | K | e | K |/|/              | f | f | b |/|/              | K | f | K |/|/
   :-----------: f/              :-----------: f/              :-----------: f/
   | e | e | e |/ e,K & r,f      | e | e | e |/ r,b            | e | e | e |/ r,b
   '-----------' inverted        '-----------' READY           '-----------' done

S21(tv)  .-----------,          S22(tv)  .-----------,
        / b / t / t /|                  / b / t / t /|
       /---/---/---/ r:                /---/---/---/ r:
      / b / t / t /|/|                / b / t / t /|/|
     /---/---/---/ r/ K:-K,b      /---/---/---/ r/ r:
    / b / t / t /|/|/|            / b / t / t /|/|/|
   :-----------: r/ f/ r:        :-----------: r/ f/ r:
   | K | f | f |/|/|/|           | K | f | f |/|/|/|
   :-----------: K/ r/            :-----------: e/ r/
   | e | e | e |/|/          f,b-| b | e | f |/|/
   :-----------: r/              :-----------: r/
   | K | f | f |/ K,b            | K | f | f |/ K,b done
   '-----------' READY           '-----------' f,b READY

S21(bv)  .-----------,          S22(bv)  .-----------,
        / b / b / b /|                  / b / b / b /|
       /---/---/---/ K:                /---/---/---/ K:
      / r / b / b /|/|        f,r-/ r / b / b /|/|
     /---/---/---/ f/ b:-K,b      /---/---/---/ K/ K:
    / t / t / t /|/|/|            / t / t / t /|/|/|
   :-----------: f/ r/ K:        :-----------: f/ r/ K:
   | e | e | e |/|/|/|           | e | e | e |/|/|/|
   :-----------: r/ K/            :-----------: e/ K/
   | f | K | K |/|/          f,b-| f | K | K |/|/
   :-----------: f/              :-----------: f/
   | e | e | e |/ K,b            | e | e | e |/ K,b done
   '-----------' READY           '-----------' f,b READY
```

Figure S21 shows that after the middle layer is now adjusted 2Qt's to the right, the K,b edge cube in the middle layer is the one and only one in a READY position to bump a cube in the bottom layer (which happens to be the f,b edge cube) by a R/V/1/1 translation on the right 'r' face (front 'f' center cube towards you). Figure S22 shows the CUBE after the second of these three bumping operations. However, now no adjustment is necessary, so the f,b edge cube in the middle layer is the one and only one in a READY position to bump the last edge cube in the bottom layer (which happens to be the f,r edge cube) by a R/V/1/1 translation on the left 'e' face (back 'K' center cube towards you). Figure S23 shows the CUBE after the third of these three bumping operations. Always adjust the middle layer first (middle layer 1Qt to the left). Then restore

```
S23(tv)  .-----------,
        / b / t / t /|
       /---/---/---/ r :
      / b / t / t /|/|
     /---/---/---/ r / f :
    / b / t / t /|/|/|
   :-----------: r / f / r :
   | k | f | f |/|/|/|/
   :-----------: k / r /
   | e | e | e |/|/|/
   :-----------: r /
   | k | f | f |/ f,b
   '-----------'  done

S23(bv)  .-----------,
        / b / b / b /|
       /---/---/---/ k :
      / b / b / b /|/|
     /---/---/---/ k / r :
    / t / t / t /|/|/|
   :-----------: f / r / k :
   | e | e | e |/|/|/|/
   :-----------: r / k /
   | f | k | k |/|/|/
   :-----------: f /
   | e | e | e |/ f,b
   '-----------'  done
```

```
S24(tv)  .-----------,
        / t / t / t /|
       /---/---/---/ r :
      / t / t / t /|/|
     /---/---/---/ r / r :
    / t / t / t /|/|/|
   :-----------: r / r / r :
   | f | f | f |/|/|/|/
   :-----------: r / r /
   | f | f | f |/|/|/
   :-----------: r /
   | f | f | f |/
   '-----------'  DONE!

S24(bv)  .-----------,
        / b / b / b /|
       /---/---/---/ k :
      / b / b / b /|/|
     /---/---/---/ k / k :
    / b / b / b /|/|/|
   :-----------: k / k / k :
   | e | e | e |/|/|/|/
   :-----------: k / k /
   | e | e | e |/|/|/
   :-----------: k /
   | e | e | e |/
   '-----------'  DONE!
```

the left 'e' face 2Qt's. Figure S24 shows the pristine CUBE after these two adjustments. Well, the CUBE works and the CUBIK solution works, so let's do it again! (IT IS PRUDENT AFTER THIS EXORCISE THAT I ADD AN 'OUT-OF-SYNC' SUBROUTINE TO MY FLOWCHART.) Refer to Table S for the brief summary of Exorcise 1 that follows.

EXORCISE-ONE SUMMARY

a) Four of the eight top face cubes are okay after 16 steps move them from the side of either the middle or bottom layer.

b) The other four top face cubes are okay after seven steps move two edge cubes whose 't' face was on the bottom face and sixteen steps for moving-out and then moving two corner cubes whose 't' face was also on the bottom face.

c) The bottom corner cubes are positioned correctly after a RR/O operation Interchanges diagonally two of the corners. (Table 10EX at the end of this 'exorcise' section shows that this diagonal interchange only occurs once in these ten exercises.)

d) A R/B/1/1 translation puts the bottom corner cubes 'in-sync'. Then two subsequent R/B/1/1 translations finish the bottom four-corner cubes. The bottom 'b' four-corner cubes are now okay.

e) The four bottom 'b' edge cubes are okay after four R/V/1/1 translations moves them to the bottom layer after they are 'adjusted' to a READY position. However, before the R/V/1/1 translation, a face needed to be inverted (2Qt's) in order to put two of these edge cubes in a READY position.

f) The four middle edge cubes are correctly positioned after two subsequent RM/O operations rotates twice three of the four middle edge cubes. Two of them are also oriented correctly (are okay).

g) The CUBE is done after an ALGO1 operation orients the remaining two middle edge cubes. However, before the operation a face needed to be inverted (2Qt's) in order to make these two edge cubes 'adjacent' (they were on a diagonal) for an ALGO1 READY position. (Table 10EX at the end of this 'exercise' section shows that this diagonal re-orientation occurs twice in these ten exercises.)

NOTES:

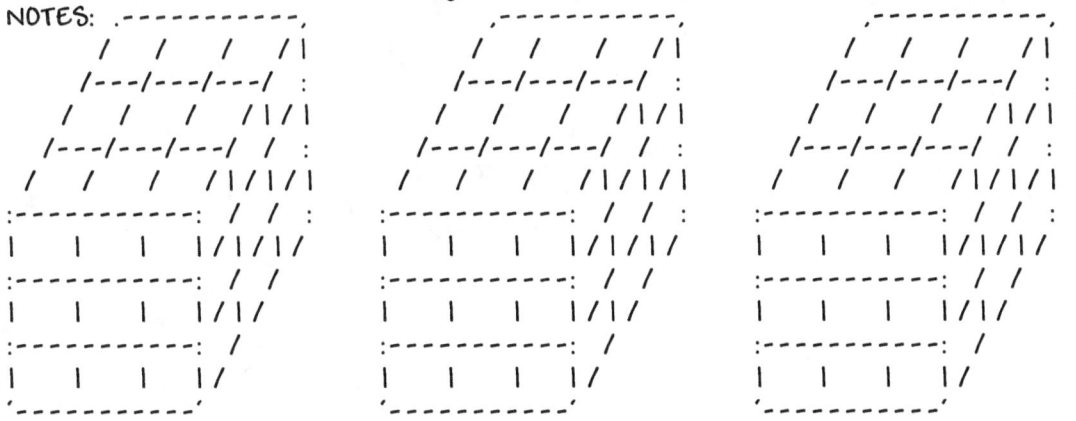

TABLE S   EXORCISE-ONE

OPER-ATION	STEPS	TWISTS	(Bottom Corners Cont'd)			(DO MIDDLE EDGES)		
			RR/0	10	1	RM/0	4	2
----------------			R/B/1/1	8	--	RM/0	4	2
t,f,e 't' cube	4	--	R/B/1/1	8	--	Invert 'e'	1	--
t,k,e 't' cube	4'	--	R/B/1/1	8	--	(ALG01)		
t,e 't' cube	4	1	-------------------			R/V/1/1	7	3
t,k 't' cube	4'	--	Corners Done	35	36	R/V/1/1	7	3
Align	1'	--	==================			C/B/1/2	7	4
t,f 't' cube	4	1	(DO BOTTOM EDGES)			Adjust(-1Qt)	1	1
t,r 't' cube	3	1	Invert 'f'	1	--	R/V/1/1	7	3
(t,f,r) 't' cube	4	--	R/V/1/1	7	3	Adjust(+2Qt)	1	1
t,f,r 't' cube	4	--	Re-invert	1	--	R/V/1/1	7	3
(t,k,r) 't' cube	4	--	Invert 'r'	1	--	R/V/1/1	7	3
----------------			R/V/1/1	7	3	Align	1	1
Top Done	40	43	Re-invert	1	--	(ALG01 Done	45	67)
================			R/V/1/1	7	3	Re-invert 'e'	1	--
(DO BOTTOM CORNERS)			Adjust	1	1	-------------------		
Align	1	--	R/V/1/1	7	3	Edges Done	55	81
(reference cube zero)			Align	1	1	==================		
			-------------------			CUBE DONE	164	208
			Bottom Done	34	48	==================		
			==================					

TABLE S   NOTES

STEPS: A step is one twist of the CUBE that can be one 'Q' turn or two 'Q' turns, e.g. inverting a face is two 'Q' turns.

TWISTS: A twist column shows when a 'step' is two twists instead of one twist, e.g. when bringing down the center slice along with the right side, an extra twist restores the right side; similarly, when rotating the middle slice (layer) with the bottom layer, an extra twist restores the bottom layer. (The vertical slice is not used in solving the CUBE, but is used in pattern generations from a solid CUBE.)

' (prime): Sometimes at the beginning when doing the top 't' face, I move the bottom layer and middle layer together.

ALIGN: An alignment step aligns the faces of the CUBE (a correction).

(xxx): A top 't' layer cube that requires two operations since it is either on the bottom 'b' face or incorrectly in the top 't' layer.

ADJUST: A step that places an edge cube, that is to be sent to the bottom 'b' layer, in a READY position.

```
S1'(tv)   .------------,
        / f / b / e / |
       /---/---/---/ b :
  t,f-/ t / t / r / |/ |
     /---/---/---/ f / e :
    / k / b / f / |/ |/ |
   :-----------: r / r / b :
   | t | k | b |/ |/ |/ |
   :-----------: e / r /
   | t | f | k |/ |/
   :-----------: r /
   | e | k | k |/  TWENTY
   '-----------'   TWISTS

S1'(bv)   .------------,
        / t / t / k / |
       /---/---/---/ r :
    / t / b / f / |/ |
   /---/---/---/ e / b :
  / b / r / f / |/ |/ |
 :-----------: t / k / f :
 | k | k | r |/ |/ |/ |
 :-----------: b / r /
 | e | e | f |/ |/  #=f,e,b
 :-----------: t /
 | e | f | e |/  TWENTY
 '-----------'   TWISTS

S3'(tv)   .------------,
        / t / t / t / |
       /---/---/---/ r :
    / t / t / t / |/ |
   /---/---/---/ r / b :-e,b
  / t / t / t / |/ |/ |
 :-----------: r / r / k :
 | f | f | f |/ |/ |/ |
 :-----------: k / e /
 | k | f | r |/ |/
 :-----------: b /
 | f | f | r |/  cor.
 '-----------'   pos. done

S3'(bv)   .------------,
        / f / k / r / |
       /---/---/---/ b :
    / e / b / f / |/ |
   /---/---/---/ r / e :-e,b
  / b / f / k /|/ |/ |
 :-----------: e / k / k :
 | e | b | b |/ |/ |/ |
 :-----------: r / k /
 | b | e | b |/ |/
 :-----------: k /
 | e | e | e |/  cor.
 '-----------'   pos. done
```

```
S2'(tv)   .------------,
        / t / t / t / |
       /---/---/---/ r :
    / t / t / t / |/ |
   /---/---/---/ r / b :
  / t / t / t / |/ |/ |
 :-----------: r / r / b :
 | f | f | f |/ |/ |/ |
 :-----------: k / b /
 | k | f | r |/ |/
 :-----------: b /
#-| e | e | k |/  top
 '-----------'   done

S2'(bv)   .------------,
        / r / f / e / |
       /---/---/---/ k :
    / k / b / r / |/ |
   /---/---/---/ f / e :
  / f / f / f / |/ |/ |
#-:-----------: r / k / k :
 | b | e | b |/ |/ |/ |
 :-----------: r / k /
 | b | e | b |/ |/
 :-----------: k /
 | e | e | e |/  top
 '-----------'   done

S4'(tv)   .------------,
        / b / k / b / |
       /---/---/---/ e :
    / t / t / t / |/ |
   /---/---/---/ r / b :
  / t / t / t / |/ |/ |
 :-----------: r / r / e :
 | f | f | f |/ |/ |/ |
 :-----------: k / f /
 | k | f | r |/ |/
 :-----------: r / cor's
 | f | r | f |/  done,
 '-----------'   e,b ready

S4'(bv)   .------------,
        / b / e / t / |
       /---/---/---/ k :
    / f / b / t / |/ |
   /---/---/---/ k / r :
  / b / f / t / |/ |/ |
 :-----------: k / k / k :
 | e | b | r |/ |/ |/ |
 :-----------: e / e /
 | b | e | b |/ |/
 :-----------: k / cor's
 | e | e | r |/  done
 '-----------'   e,b ready
```

EXAMPLE #2 LET'S DO IT AGAIN!!

Figure S1' is a scrambled CUBE (after 20 twists with eyes closed). Figure S2' is the CUBE after moving seven top 't' face cubes that were on the sides of either the middle layer or the bottom layer and one corner cube in the top layer as follows: Top layer 1Qt to the right aligns the t,f cube (see Figure S1'). Cube t,r,f is done per Figure A of the FOUR STEP APPROACH REVIEW.

Cube t,e is done per Figure E. Cube t,r is also done per Figure E. After aligning the faces, cube t,e,f is done per Figure B. Cube t,r,k is done per 2Qt's (two steps moved the corner cube to the top without having to dip the top down). Cube t,k is done per Figure G. Cube t,e,k is done per Figure H which moves it out of the top layer so that Figure A finishes it. The top is done. It took 31 steps to finish the top as shown in Figure S2'. Perusal of Figure S2' will show that corner cube f,e,b is the only corner positioned correctly (not oriented), therefore, rotating the layer 1Qt to the right causes corner cubes k,r,b and k,e,b to be positioned correctly (not oriented). Hence, a R/0 operation on the back 'k' face interchanges the other two corner cubes thereby positioning all four bottom corner cubes correctly as shown in Figure S3'. Lo and behold, the corner cube f,e,b is correctly oriented and the other three corner cubes are already 'in-sync'. So two R/B/1/1 translations on the front 'f' face finishes the bottom four corners as shown in Figure S4'. It took only 26 steps to finish the corners, WOW. Perusal of Figure S3' shows that by inverting the back 'k' face, the e,b edge cube is in a ready position to bump back the f,b edge cube in the bottom layer. Figure S4' shows the e,b edge cube in the READY position. Figure S5' shows the

```
S5'(tv)  .-----------,              S6'(tv)  ,-----------,
        / t / t / t / |                     / t / t / t / |
       /---/---/---/ r :                   /---/---/---/ r :
      / t / t / t / | / |                 / t / t / t / | / |
     /---/---/---/ r / k :-k,b           /---/---/---/ r / k :
    / t / t / t / | / | / |             / t / t / t / | / | / |
   :-----------: r / r / r:            :-----------: r / r / r:
   | f | f | f | f |/ | / |            | f | f | f |/ | / | / |
   :-----------: f / f /               :-----------: r / f /
   | r | f | b |/ | / /                | k | f | b |/ | / /
   :-----------: r /                   :-----------: r /
   | f | r | f |/ e,b done,            | f | r | f |/  k,b
   '-----------'  k,b ready            '-----------'  done

S5'(bv)  .-----------,              S6'(bv)  ,-----------,
        / b / e / b / |                     / b / e / b / |
       /---/---/---/ k :                   /---/---/---/ k :
      / f / b / k / | / |           r,f-/ f / b / b / | / |
     /---/---/---/ e / b :-k,b           /---/---/---/ k / r :
    / b / b / b / | / | / |             / b / b / b / | / | / |
   :-----------: k / k / k:            :-----------: k / k / k:
   | e | e | e | e |/ | / |            | e | e | e |/ | / | / |
   :-----------: k / k /               :-----------: b / k /
   | b | e | r |/ | / /                | e | e | f |/ | / /
   :-----------: k /                   :-----------: k /
   | e | e | e | e |/ e,b done,        | e | e | e |/  k,b
   '-----------'  k,b ready            '-----------'  done

S7'(tv)  .-----------,              S8'(tv)  ,-----------,
        / t / t / t / |                     / t / t / t / |
       /---/---/---/ r :                   /---/---/---/ r :
      / t / t / t / | / |                 / t / t / t / | / |
     /---/---/---/ r / r :             /---/---/---/ r / r :
    / t / t / t / | / | / |             / t / t / t / | / | / |
   :-----------: r / r / r:            :-----------: r / e / r:
   | f | f | f | f |/ | / |            | f | f | f | f |/ | / |
   :-----------: e / f / -e,f          :-----------: b / f / -e,f
   | k | f | k | k |/ | /              | f | k | r |/ | / /
   :-----------: r /                   :-----------: r /
   | f | f | f |/ f,b                  | f | f | f |/  r,b
   '-----------'  done                 '-----------'  ready

S7'(bv)  .-----------,              S8'(bv)  ,-----------,
        / b / e / b / |                     / b / e / b / |
       /---/---/---/ k :                   /---/---/---/ k :
  f,b-/ b / b / b / | / |                 / b / b / b / | / |
     /---/---/---/ k / f :             /---/---/---/ k / k :
    / b / b / b / | / | / |             / b / b / b / | / | / |
   :-----------: k / k / k:            :-----------: k / f / k:
   | e | e | e | e |/ | / |            | e | e | e |/ | / | / |
   :-----------: r / k /               :-----------: k / k /
   | r | e | b |/ | / /                | r | r | e |/ | / /
   :-----------: k /                   :-----------: k /
   | e | e | e | e |/ f,b              | e | e | e |/  r,b
   '-----------'  done                 '-----------'  ready
```

CUBE after a R/V/1/1 translation on the back 'k' face sends the e,b edge cube to the bottom layer and the back 'k' face is re-inverted. Perusal of Figure S5' shows that the k,b edge cube is already in a READY position to bump back the e,k edge cube. Figure S6' shows the CUBE after a R/V/1/1 translation on the right 'r' face sends the k,b edge cube to the bottom layer. Similar to the e,b edge cube, inverting the left 'e' face puts the f,b edge cube in a READY position to bump back the r,f edge cube. Figure S7' shows the CUBE after inverting the left 'e' face and doing a R/V/1/1 translation on the left 'e' face which sends the f,b edge cube to the bottom layer then re-inverting the left face. Figure S8' shows the CUBE after the middle layer is adjusted 2Qt's putting the r,b edge cube in a READY position to bump back the e,f edge cube. Figure S9' shows the CUBE after a R/V/1/1 translation on the front 'f face sends the r,b edge cube to the bottom layer which completes the bottom. FOUR MORE CUBES TO GO! Figure S10' shows the CUBE after it is re-aligned (Middle layer 2Qt's). Perusal of Figure S10' shows that a C/V/1/2 translation interchanges the middle edge cubes along the sides correctly positions them. Figure S11' shows the CUBE after inverting the right 'r' face (otherwise they are interchanged diagonally instead) and doing a C/V/1/2 translation on the right 'r' face. Perusal of Figure S11' shows middle edge cubes f,e and f,r correctly oriented but cubes k,e and k,r need to be oriented by ALGO1 lwhich finishes the CUBE. Figure S12' shows the CUBE after two R/V/1/1 translations (on the back 'k face then the left 'e' face) sends the k,r and k,e edge cubes to the bottom layer so they are READY for a C/B/1/2 translation to invert them.

```
S9'(tv)    .-----------,              S10'(tv)   .-----------,              S11'(tv)   .-----------,
         / t / t / t /1                       / t / t / t /1                        / t / t / t /1
        /---/---/---/ r:                      /---/---/---/ r:                      /---/---/---/ r:
        / t / t / t /1/1                      / t / t / t /1/1                      / t / t / t /1/1
       /---/---/---/ r / r:                   /---/---/---/ r / k:                  /---/---/---/ r / k:-k,r
       / t / t / t /1/1/1                     / t / t / t /1/1/1                    / t / t / t /1/1/1
      :-----------: r / e / r:                :-----------: r / r / r:             :-----------: r / r / r:
      | f | f | f | f /1/1/                   | f | f | f | f /1/1/                 | f | f | f | f /1/1/
      :-----------: k / r /-r,b               :-----------: e / r /                :-----------: r / r /
      | e | k | r |1/1/                       | f | f | f | f /1/                   | f | f | f | f /1/
      :-----------: r /                       :-----------: r /                     :-----------: r /
      | f | f | f | f /  r,b                  | f | f | f | f / bottom             | f | f | f | f / middle
      '-----------' done                      '-----------' done                  '-----------' pos. done

S9'(bv)    .-----------,              S10'(bv)   .-----------,              S11'(bv)   .-----------,
         / b / b / b /1                       / b / b / b /1                        / b / b / b /1
        /---/---/---/ k:                      /---/---/---/ k:                      /---/---/---/ k:
        / b / b / b /1/1                      / b / b / b /1/1                      / b / b / b /1/1
       /---/---/---/ k / f:                   /---/---/---/ k / e:                  /---/---/---/ k / r:-k,r
       / b / b / b /1/1/1                     / b / b / b /1/1/1                    / b / b / b /1/1/1
      :-----------: k / f / k:                :-----------: k / k / k:             :-----------: k / k / k:
      | e | e | e | e /1/1/                   | e | e | e | e /1/1/                 | e | e | e | e /1/1/
      :-----------: f / k /                   :-----------: r / k /                 :-----------: e / k /
      | k | r | e |1/1/                       | r | e | k |1/1/                     | e | e | k |1/1/
      :-----------: k /                       :-----------: k /                     :-----------: k /
      | e | e | e | e /  r,b                  | e | e | e | e / bottom             | e | e | e | e / middle
      '-----------' done                      '-----------' done                  '-----------' pos. done

S12'(tv)   .-----------,              S13'(tv)   .-----------,
         / t / t / t /1                       / t / t / t /1
        /---/---/---/ r:                      /---/---/---/ r:
        / t / t / t /1/1                      / t / t / t /1/1
       /---/---/---/ r / f:                   /---/---/---/ r / f:
       / t / t / t /1/1/1                     / t / t / t /1/1/1
      :-----------: r / r / r:                :-----------: r / r / r:
      | f | f | f | f /1/1/                   | f | f | f | f /1/1/
      :-----------: f / r /                   :-----------: f / r /
      | b | f | e |1/1/                       | b | f | e |1/1/
      :-----------: r /                       :-----------: r /
      | f | k | f | f / k,r & k,e             | f | k | f | f / k,r & k,e
      '-----------' on bottom                 '-----------' inverted

S12'(bv)   .-----------,              S13'(bv)   .-----------,
         / b / b / b /1                       / b / b / b /1
        /---/---/---/ k:                      /---/---/---/ k:
     k,r-/ r / b / b /1/1                     / b / b / e /1/1
        /---/---/---/ k / b:                  /---/---/---/ k / b:
        / b / k / b /1/1/1                    / b / k / b /1/1/1
      :-----------: k / k / k:                :-----------: k / k / k:
      | e | e | e | e /1/1/                   | e | r | e | e /1/1/
      :-----------: f / k /                   :-----------: f / k /
      | e | e | r |1/1/                       | e | e | r |1/1/
      :-----------: k /                       :-----------: k /
      | e | e | e | e / k,r & k,e             | e | e | e | e / k,r & k,e
      '-----------' on bottom                 '-----------' inverted
```

Figure S13' shows the CUBE after they sre inverted by a C/B/1/2 translation on the right 'r' face. Figure S14' shows the CUBE, (1) after a R/V/1/1 translation with a -Qt READY adjustment bumps back the k,r edge cube with the b,e edge cube, (2) after a R/V/1/1 translation with a +2Qt's READY adjustment bumps back the k,b edge cube with the f,b edge cube, and (3) after a R/V/1/1 translation with no READY adjustment bumps back the k,e edge cube with the k,b edge cube which finishes all the translations in ALGO1. A perusal of Figure S14' shows that rotating the mid-dle layer 1Qt to the left results in a pristine configuration shown in Figure S15'. WOW! Only 146 steps (192 twists) where as the first exorcise required 164 steps (208 twists). What fun, ONE-MORE-TIME!!! Refer to Table S' for the brief summary of Exorcise 2 that follows.

```
S14'(tv) .-----------,
        / t / t / t /|
       /---/---/---/ r :
       / t / t / t /|/|
      /---/---/---/ r / f :
      / t / t / t /|/|/|
     :-----------: r / f / r :
     | f | f | f |/|/|/|/
     :-----------: f / r /
     | e | e | e |/|/|/
     :-----------: r /
     | f | f | f |/ ALG01
     '-----------' ending
```

```
S15'(tv) .-----------,
        / t / t / t /|
       /---/---/---/ r :
       / t / t / t /|/|
      /---/---/---/ r / r :
      / t / t / t /|/|/|
     :-----------: r / r / r :
     | f | f | f |/|/|/|/
     :-----------: r / r /
     | f | f | f |/|/|/
     :-----------: r /
     | f | f | f |/ CUBE
     '-----------' DONE
```

```
S14'(bv) .-----------,
        / b / b / b /|
       /---/---/---/ k :
       / k / b / b /|/|
      /---/---/---/ k / r :
      / b / b / b /|/|/|
     :-----------: k / r / k :
     | e | e | e |/|/|/|/
     :-----------: r / k /
     | k | k | k |/|/|/
     :-----------: k /
     | e | e | e |/ ALG01
     '-----------' ending
```

```
S15'(bv) .-----------,
        / b / b / b /|
       /---/---/---/ k :
       / b / b / b /|/|
      /---/---/---/ k / k :
      / b / b / b /|/|/|
     :-----------: k / k / k :
     | e | e | e |/|/|/|/
     :-----------: k / k /
     | e | e | e |/|/|/
     :-----------: k /
     | e | e | e |/ CUBE
     '-----------' DONE
```

<u>EXORCISE-TWO SUMMARY</u>

a) Six of the eight top layer cubes are okay after 19 steps move them from the sides of either the middle layer or the bottom layer.

b) The other two top layer cubes are okay after four steps move an edge cube whose top 't' face was on the bottom face and seven steps for moving-down and then move a corner cube that was in the top layer incorrectly. The top layer is now done.

c) Since at least two bottom corner cubes must be correctly positioned (either adjacent or diagonally) the bottom layer is further rotated until two correctly positioned bottom corners are found. Then all four bottom corner cubes are positioned correctly after a R/O slice operation interchanges two adjacent corner cubes.

d) By happenstance one bottom corner is okay and the other three are already 'in-sync'. Hence, two subsequent R/B/I/I translations finishes the bottom corners (they are all okay).

e) The four bottom edge cubes are okay after four R/V/I/I translations move them to the bottom layer after they are 'adjusted' to a READY position. However, before the R/V/I/I translation a face needed to be inverted (2Qt's) in order to put two of these four edge cubes in a READY position.

f) The four middle edge cubes are correctly positioned after a C/V/I/2 translation interchanges them along the sides. However, before the C/V/I/2 translation a face needed to be inverted (2Qt's), otherwise, this translation would have interchanged them diagonally instead. (Table 10EX at the end of this exorcise section shows that this interchange along the sides only occurs once in these ten exercises. For Exorcise 8 they were interchanged diagonally.) Two of the middle edge cubes end up okay after the interchange.

g) The CUBE is done after a ALG01 operation orients the two remaining adjacent middle edge cubes.

NOTES:

```
.-----------,      .-----------,      .-----------,
/ / / /|       / / / /|       / / / /|
/---/---/---/ :   /---/---/---/ :   /---/---/---/ :
/ / / /|/|      / / / /|/|      / / / /|/|
/---/---/---/ / :  /---/---/---/ / :  /---/---/---/ / :
/ / / /|/|/|     / / / /|/|/|     / / / /|/|/|
:-----------: / / :  :-----------: / / :  :-----------: / / :
| | | |/|/|/|    | | | |/|/|/|    | | | |/|/|/|
:-----------: / /   :-----------: / /   :-----------: / /
| | | |/|/|     | | | |/|/|     | | | |/|/|
:-----------: /    :-----------: /    :-----------: /
| | | |/|      | | | |/|      | | | |/|
'-----------'      '-----------'      '-----------'
```

TABLE S'   EXORCISE-TWO

OPER-ATION	STEPS/TWS	(#) FIG.	(DO BOTTOM EDGES)			(DO MIDDLE EDGES)		
			Invert 'k'	1 / --		Invert 'r'	1 / --	
			R/V/1/1	7 / 3		C/V/1/2	8 / 8	
t,f 't' cube	1 / --	+1Qt	Re-invert	1 / --		Align	1 / --	
t,r,f 't' cube	4 / --	A	R/V/1/1	7 / 3		(ALGO1)		
t,e 't' cube	4' / --	E	Invert 'e'	1 / --		R/V/1/1	7 / 3	
t,r 't' cube	4' / --	E	R/V/1/1	7 / 3		R/V/1/1	7 / 3	
Align	1 / --	Qt's	Re-invert	1 / --		C/B/1/2	7 / 4	
t,e,f 't' cube	4 / --	B	Adjust	1 / 1		Adjust(-1Qt)	1 / 1	
t,r,k 't' cube	2 / --	--	R/V/1/1	7 / 3		R/V/1/1	7 / 3	
t,k 't' cube	4 / 2	G	Align	1 / 1		Adjust(+2Qt)	1 / 1	
(t,e,k) 't' cube	3 / --	H	------------------			R/V/1/1	7 / 3	
t,e,k 't' cube	4 / --	A	Bottom Done 34 / 48			R/V/1/1	7 / 3	
------------------			==================			Align	1 / 1	
Top Done	31 / 33		NOTE:			(ALGO1 Done 45 / 67)		
==================			(#) See FOUR STEP			------------------		
(DO BOTTOM CORNERS)			APPROACH REVIEW for			Edges Done 55 / 85		
Align	1 / --		Figures A through Figure H.			==================		
R/0	9 / --					CUBE DONE 146 / 192		
R/B/1/1	8 / --					==================		
R/B/1/1	8 / --							

Corners Done 26 / 26								
==================								

TABLE S'   NOTES

STEPS: A step is one twist of the CUBE that can be one 'Q' turn or two 'Q' turns, e.g. inverting a face is two 'Q' turns.

TWISTS: A twist column shows when a 'step' is two twists instead of one twist, e.g. when bringing down the center slice along with the right side, an extra twist restores the right side; similarly, when rotating the middle slice (layer) with the bottom layer, an extra twist restores the bottom layer. (The vertical slice is not used in solving the CUBE, but is used in pattern generations from a solid CUBE.)

' (prime): Sometimes at the beginning when doing the top 't' face, I move the bottom layer and middle layer together.

ALIGN: An alignment step aligns the faces of the CUBE (a correction).

(xxx): A top 't' layer cube that requires two operations since it is either on the bottom 'b' face or incorrectly in the top 't' layer.

ADJUST: A step that places an edge cube, that is to be sent to the bottom 'b' layer, in a READY position.

```
S1"(tv)  .-----------,
        / e / f / k /|
       /---/---/---/ r :
      / k / t / t /|/|
     /---/---/---/ e / t:
    / f / b / k /|/|/|
   :-----------: t / r / f:
   | t | f | e |/|/|/|
   :-----------: k / r /
   | t | f | e |/|/|
   :-----------: e /
   | k | k | f |/|THIRTY
   '-----------' TWISTS

S1"(bv)  .-----------,
        / t / k / b /|
       /---/---/---/ r :
      / t / b / f /|/|
     /---/---/---/ e / f:
    / e / b / t /|/|/|
   :-----------: k / k / b:
   | b | e | r |/|/|/|
   :-----------: r / r /
   | r | e | b |/|/|
   :-----------: b /
   | r | b | f |/|THIRTY
   '-----------' TWISTS

S3"(tv)  .-----------,
        / t / t / t /|
       /---/---/---/ r :
      / t / t / t /|/|
     /---/---/---/ r / r:
    / t / t / t /|/|/|
   :-----------: r / r / b:-rkb
   | f | f | f |/|/|/|
   :-----------: e / k /
   | b | f | k |/|/|
   :-----------: e /
   | f | f | b |/| k,b,r &
   '-----------' k,b,e pos.

S3"(bv)  .-----------,
        / f / b / k /|
       /---/---/---/ r :-r,k,b
      / e / b / f /|/|
     /---/---/---/ r / k:
    / r / e / b /|/|/|
   :-----------: k / k / k: feb-
   | b | b | e |/|/|/|
   :-----------: b / k /
   | r | e | f |/|/|
   :-----------: k /
   | e | e | e |/| k,b,r &
   '-----------' k,b,e pos.
```

```
S2"(tv)  ,-----------,
        / t / t / t /|
       /---/---/---/ r :
      / t / t / t /|/|
     /---/---/---/ r / r:
    / t / t / t /|/|/|
   :-----------: r / r / b:
   | f | f | f |/|/|/|
   :-----------: e / f /
   | b | f | k |/|/|
   :-----------: f /-f,b,r
   | e | b | b |/| top
   '-----------' done

S2"(bv)  ,-----------,
     f.b.r-/ r / e / f /|
       /---/---/---/ e :
      / e / b / b /|/|
     /---/---/---/ k / k:
    / b / f / k /|/|/|
   :-----------: b / k / k:
   | k | r | r |/|/|/|
   :-----------: b / k /
   | r | e | f |/|/|
   :-----------: k /
   | e | e | e |/| top
   '-----------' done

S4"(tv)  ,-----------,
        / t / t / t /|
       /---/---/---/ r :
      / t / t / t /|/|
     /---/---/---/ r / r:
    / t / t / t /|/|/|
   :-----------: r / r / b:
   | f | f | f |/|/|/|
   :-----------: e / k /
   | b | f | k |/|/|
   :-----------: f /
   | f | e | b |/| cor.
   '-----------' pos. done

S4"(bv)  ,-----------,
        / r / b / k /|
       /---/---/---/ r :
      / f / b / b /|/|
     /---/---/---/ e / k:
    / b / f / e /|/|/|
   :-----------: b / k / k:
   | e | r | k |/|/|/|
   :-----------: b / k /
   | r | e | f |/|/|
   :-----------: k /
   | e | e | e |/| cor.
   '-----------' pos. done
```

EXAMPLE #3 ONE-MORE-TIME!!!
Figure S1" is a scrambled CUBE (after 30 twists with eyes closed). Figure S2" is the CUBE after moving seven top 't' face cubes that were on the sides of either the middle layer or the bottom layer (no top face 't' cubes ended up on the bottom nor on the top incorrectly). It took 27 steps to finish the top 't' layer as shown in Figure S2". Perusal of Figure S2" shows that corner cube f,b,r is already positioned correctly (only one though) so rotating the bottom layer 1Qt to the left positions the k,b,r and k,b,e bottom corner cubes correctly as shown in Figure S3". Figure S4" is the CUBE after doing a R/0 operation on the back 'k' face that interchanges the other two corners thereby positioning all four bottom corner cubes correctly as shown in Figure S4". Lo and behold it happened again. The corner cube f,e,b is correctly oriented and the other three bottom corner cubes are already 'in-sync', but this time, only one R/B/1/1 translation on the front 'f' face finishes the bottom four corner cubes as shown in Figure S5". Perusal of Figure S5" shows that by inverting the left 'e' face, the f,b edge cube is made READY to bump back the k,b edge cube. Figure S6" shows the f,b edge cube in the READY position. Figure S7" shows the CUBE after a R/V/1/1 translation on the left 'e' face sends the f,b edge cube to the bottom layer and the left face is re-inverted. Perusal of Figure S7" shows that the k,b edge cube is already in a READY position to bump back the f,e edge cube. Figure S8" shows the CUBE after a R/V/1/1 translation on the right 'r' face sends the k,b edge cube to the bottom layer. Figure S9" shows that edge cube r,b is made READY by a double adjustment (first the middle layer is rotated 1Qt to the right;

```
S5"(tv)   .-----------,
         / t / t / t /|
        /---/---/---/ r:
        / t / t / t /|/|
       /---/---/---/ r / r:
       / t / t / t /|/|/|
      :-----------: r / r / r:
      | f | f | f |/|/|/|
      :-----------: e / e /
      | b | f | K |/|/
      :-----------: r /
      | f | K | f |/  corn's
      '-----------'  done
```

```
S6"(tv)   .-----------,
         / b / t / t /|
        /---/---/---/ r:
        / f / t / t /|/|
       /---/---/---/ r / r:
       / b / t / t /|/|/|
      :-----------: r / r / r:
      | K | f | f |/|/|/|
      :-----------: e / e /
 f,b-| b | f | K |/|/
      :-----------: r /
      | K | K | f |/  f,b
      '-----------'  ready
```

```
S7"(tv)   .-----------,
         / t / t / t /|
        /---/---/---/ r:
        / t / t / t /|/|
       /---/---/---/ r / K :-K,b
       / t / t / t /|/|/|
      :-----------: r / r / r:
      | f | f | f |/|/|/|
      :-----------: r / e /
      | r | f | b |/|/
      :-----------: r /
      | f | f | f |/  f,b done,
      '-----------'  K,b ready
```

```
S5"(bv)   .-----------,
         / b / b / b /|
        /---/---/---/ K:
  K,b-/ b / b / f /|/|
      /---/---/---/ e / K:
      / b / f / b /|/|/|
      :-----------: K / K / K:
      | e | r | e |/|/|/|
      :-----------: b / K /
      | r | e | f |/|/
      :-----------: K /
      | e | e | e |/  corn's
      '-----------'  done
```

```
S6"(bv)   ,-----------,
         / b / b / b /|
        /---/---/---/ K:
  K,b-/ b / b / b /|/|
      /---/---/---/ e / K:
      / t / t / t /|/|/|
      :-----------: f / K / K:
      | e | e | e |/|/|/|
      :-----------: b / K /
 f,b-| f | e | r |/|/
      :-----------: f /
      | e | r | e |/  f,b
      '-----------'  ready
```

```
S7"(bv)   ,-----------,
         / b / b / b /|
        /---/---/---/ K:
  f,b-/ b / b / f /|/|
      /---/---/---/ e / b :-K,b
      / b / f / b /|/|/|
      :-----------: K / K / K:
      | e | r | e |/|/|/|
      :-----------: e / K /
      | K | e | K |/|/
      :-----------: K /
      | e | e | e |/  f,b done,
      '-----------'  K,b ready
```

```
S8"(tv)   .-----------,
         / t / t / t /|
        /---/---/---/ r:
        / t / t / t /|/|
       /---/---/---/ r / e:
       / t / t / t /|/|/|
      :-----------: r / r / r:
      | f | f | f |/|/|/|
      :-----------: r / e /
      | f | f | K |/|/
      :-----------: r /
      | f | f | f |/  K,b
      '-----------'  done
```

```
S9"(tv)   ,-----------,
         / t / t / t /|
        /---/---/---/ r:
        / t / t / t /|/|
       /---/---/---/ r / K:
       / b / b / b /|/|/|
      :-----------: e / f / r:
      | f | f | f |/|/|/|
      :-----------: b / e /-e,b
      | e | e | r |/|/
      :-----------: e /
      | f | f | f |/  r,b
      '-----------'  ready
```

```
S8"(bv)   .-----------,
         / b / b / b /|
        /---/---/---/ K:
        / b / b / b /|/|
       /---/---/---/ K / K:
       / b / f / b /|/|/|
      :-----------: K / K / K:
      | e | r | e |/|/|/|
      :-----------: b / K /
      | e | e | r |/|/
      :-----------: K /
      | e | e | e |/  K,b
      '-----------'  done
```

```
S9"(bv)   ,-----------,
         / t / b / b /|
        /---/---/---/ K:
        / t / b / b /|/|
       /---/---/---/ K / r:
       / t / f / b /|/|/|
      :-----------: K / r / K:
      | r | r | e |/|/|/|
      :-----------: e / K /
      | f | K | K |/|/
      :-----------: K /
      | r | e | e |/  r,b
      '-----------'  ready
```

second the front 'f' face is inverted) in order to bump back the e,b edge cube. Figure S10" shows the CUBE after a R/V/I/I translation on the front 'f' face (left 'e' center cube towards you) sends the r,b edge cube to the bottom layer and the front face is re-inverted. Perusal of S10" shows that the e,b edge cube is already in a READY position to bump back the f,r edge cube. Figure S11" shows the CUBE after a R/V/I/I translation on the back 'K' face (right 'r' center cube towards you) sends the e,b edge cube to the bottom layer and finishes the bottom layer. FOUR MIDDLE EDGE CUBES TO GO. Figure S12" aligns the middle layer via a IQt to the left. Perusal of Figure S12" shows that the K,r middle edge cube is OK (correctly positioned and oriented) but the other three middle edge cubes are positioned incorrectly.

```
S10"(tv) ,-----------,
        / t / t / t /|
       /---/---/---/ r :
      / t / t / t /|/|
     /---/---/---/ r / f:
    / t / t / t /|/|/|
   :-----------: r / f / r:
   | f | f | f |/|/|/
   :-----------: e / r /-r,b
   | K | e | K |/|/
   :-----------: r /
   | f | f | f |/ r,b done,
   '-----------' e,b ready
S10"(bv) ,-----------,
        / b / b / b /|
       /---/---/---/ K :
      / b / b / b /|/|
     /---/---/---/ K / e:
    / b / f / b /|/|/|
   :-----------: K / r / K:
   | e | r | e |/|/|/
   :-----------: e / K /
   | r | K | b |/|/
   :-----------: K /
   | e | e | e |/ r,b done,
   '-----------' e,b ready
S13"(tv) ,-----------,
        / t / t / t /|
       /---/---/---/ r :
      / t / t / t /|/|
     /---/---/---/ r / r:-k,r
    / t / t / t /|/|/|
   :-----------: r / r / r:
   | f | f | f |/|/|/
   :-----------: e / r /
   | K | f | f |/|/
   :-----------: r /
   | f | f | f |/ first
   '-----------' RM/o done
S13"(bv) ,-----------,
        / b / b / b /|
       /---/---/---/ K :
      / b / b / b /|/|
     /---/---/---/ K / K :-k,r
    / b / b / b /|/|/|
   :-----------: K / K / K:
   | e | e | e |/|/|/
   :-----------: f / K /
   | e | e | r |/|/
   :-----------: K /
   | e | e | e |/ first
   '-----------' RM/o done
```

```
S11"(tv) ,-----------,
        / t / t / t /|
       /---/---/---/ r :
      / t / t / t /|/|
     /---/---/---/ r / K:
    / t / t / t /|/|/|
   :-----------: r / f / r:
   | f | f | f |/|/|/
   :-----------: f / r /
   | e | e | r |/|/
   :-----------: r / e,b done,
   | f | f | f |/ bottom
   '-----------' done
S11"(bv) ,-----------,
        / b / b / b /|
       /---/---/---/ K :
      / b / b / b /|/|
     /---/---/---/ K / e:
    / b / b / b /|/|/|
   :-----------: K / r / K:
   | e | e | e |/|/|/
   :-----------: r / K /
   | f | K | K |/|/
   :-----------: K / e,b done
   | e | e | e |/ bottom
   '-----------' done
S14"(tv) ,-----------,
        / t / t / t /|
       /---/---/---/ r :
      / t / t / t /|/|
     /---/---/---/ r / r:
    / t / t / t /|/|/|
   :-----------: r / r / r:
   | f | f | f |/|/|/
   :-----------: r / r /
   | f | f | f |/|/
   :-----------: r / second
   | f | f | f |/ RM/o done,
   '-----------' CUBE DONE
S14"(bv) ,-----------,
        / b / b / b /|
       /---/---/---/ K :
      / b / b / b /|/|
     /---/---/---/ K / K:
    / b / b / b /|/|/|
   :-----------: K / K / K:
   | e | e | e |/|/|/
   :-----------: K / K /
   | e | e | e |/|/
   :-----------: K / second
   | e | e | e |/ RM/o done,
   '-----------' CUBE DONE
```

```
S12"(tv) ,-----------,
        / t / t / t /|
       /---/---/---/ r :
      / t / t / t /|/|
     /---/---/---/ r / r:-k,r
    / t / t / t /|/|/|
   :-----------: r / r / r:
   | f | f | f |/|/|/
   :-----------: e / r /
   | f | f | K |/|/
   :-----------: r /
   | f | f | f |/ align,
   '-----------' k,r OK
S12"(bv) ,-----------,
        / b / b / b /|
       /---/---/---/ K :
      / b / b / b /|/|
     /---/---/---/ K / K:-k,r
    / b / b / b /|/|/|
   :-----------: K / K / K:
   | e | e | e |/|/|/
   :-----------: f / K /
   | r | e | e |/|/
   :-----------: K /
   | e | e | e |/ align,
   '-----------' k,r OK
```

A RM/o operation on the back 'k' face rotates these three edge cubes until they are correctly positioned. Figure S13" shows the CUBE after the first RM/o operation on the back 'k' face. A second RM/o operation on the back 'k' face not only correctly positions the other three edge cubes but by happenstance also correctly orients them as shown in Figure S14" which finishes the CUBE. Lucky, lucky, lucky!!! Only about half the steps of the two previous exercises, 77 steps versus 146 or 164.

P.S. I never-never-never would have thought that I could have solved the CUBE in much less than 150 steps as so happened in this exercise (EX. 3). It's like winning the lottery, so why not? Exercise four is the POSTSCRIPT.

Refer to Table S" for the brief summary of Exercise 3 that follows.

EXERCISE-THREE SUMMARY

a) Seven of the eight top layer cubes are okay after 26 steps moves them from the side of either the middle layer or the bottom layer. One edge cube in the top layer was already made okay.

b) None of the top 't' layer cubes ended up incorrectly on the top or bottom layer, thus Figures F, G, H, or J of the FOUR STEP APPROACH REVIEW were not referred to.

c) Same as for Exercise 2, since at least two bottom corners must be positioned correctly, the bottom layer is further rotated until two correctly positioned corners are found. Then all four-bottom corners are positioned correctly after a R/O operation interchanges two adjacent corners.

d) Then again like Exercise 2, by happenstance one bottom corner is okay and the other three are already 'in-sync'. However, unlike Exercise 2 only one instead of two R/B/1/1 translations finishes the bottom corners (they are all okay).

e) The four bottom edge cubes are okay after four R/V/1/1 translations move them to the bottom layer after they are 'adjusted' to a READY position. However, like Exercise 1 and 2, before the R/V/1/1 translation a face needed to be inverted (2Qt's) in order to put two of these edge cubes in a READY position.

f) Like Exercise 1 the four middle edge cubes are correctly positioned after two subsequent RM/O operations rotated twice three of the four middle edge cubes. But unlike Exercise 1, all four instead of just two are correctly oriented. This finished the CUBE. (Table 10EX at the end of this exercise section shows that this immediate orientation of these four middle edge cubes occurs twice in these ten exercises.)

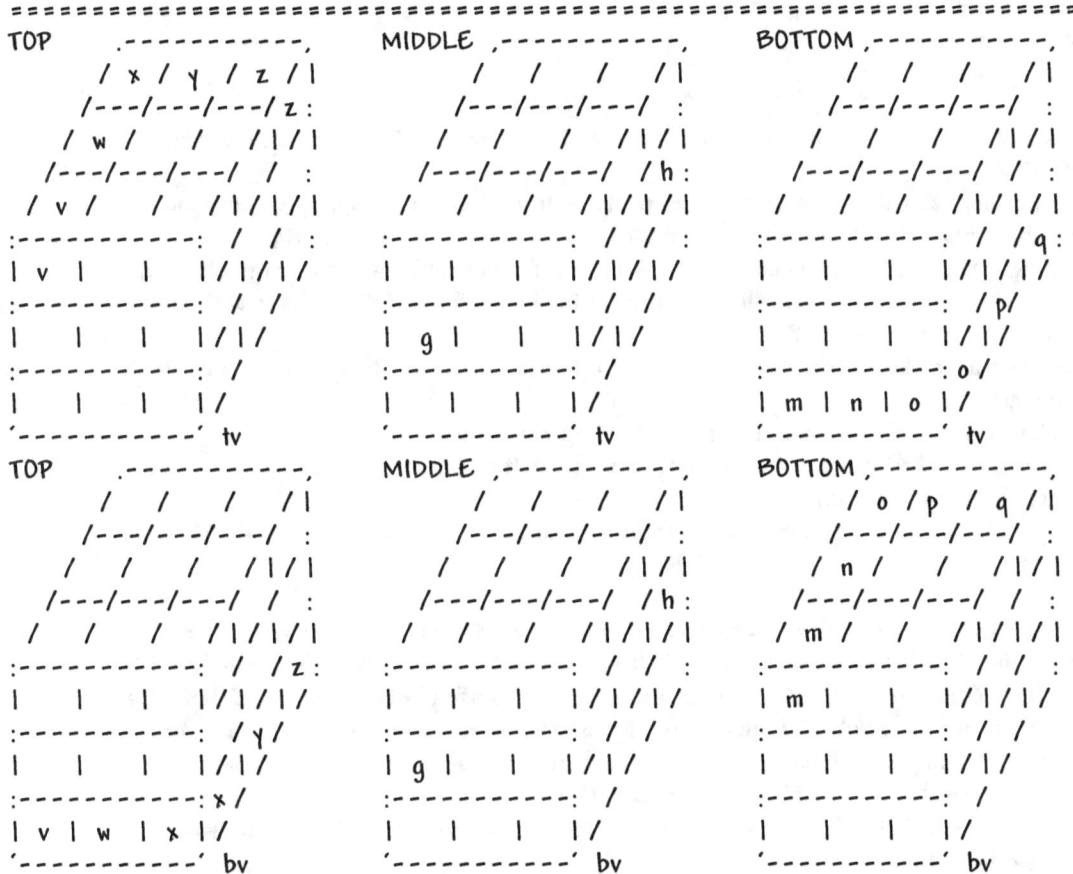

The above three sets of figures show the corresponding corner and edge cubes that share (appear) in both the top (tv) and bottom (bv) views. Of the twenty cubes (not including the six center cubes) twelve do this sharing.

TABLE S" EXORCISE-THREE

OPERATION	STEPS/TWS	(#) FIG.	(DO BOTTOM EDGES)		(DO MIDDLE EDGES)	
			Invert 'e'	1 / --	RM/O	4 / 2
			R/V/1/1	7 / 3	RM/O	4 / 2
t,e 'f' cube	1 / --	2Qt's	Re-invert	1 / --		
t,f 'f' cube	4' / --	D	R/V/1/1	7 / 3	Edges Done 8 / 12	
Align	1' / --	Qt's	Adjust (1Qt)	1 / --		
t,k 'f' cube	3' / --	E	Invert 'f'	1 / --	CUBE DONE 77 / 105	
Align	1 / --	Qt's	R/V/1/1	7 / 3		
t,e,k 'f' cube	3 / --	C	Re-invert	1 / --		
t,e,f 'f' cube	4 / --	A	R/V/1/1	7 / 3		
t,r 'f' cube	3' / --	C	Align (1Qt)	1 / 1		
Align	1 / --	Qt's				
t,r,k 'f' cube	3 / --	B	Bottom Done 34 / 48			
t,f,r 'f' cube	3 / --	B				
Top Done 27 / 27			NOTE:			

(DO BOTTOM CORNERS)

Align	1 / --
R/O	9 / --
R/B/1/1	8 / --

Corners Done 18 / 18

(#) See FOUR STEP APPROACH REVIEW for Figures A through Figure H.

TABLE S" NOTES

STEPS: A step is one twist of the CUBE that can be one 'Q' turn or two 'Q' turns, e.g. inverting a face is two 'Q' turns.

TWISTS: A twist column shows when a 'step' is two twists instead of one twist, e.g. when bringing down the center slice along with the right side, an extra twist restores the right side; similarly, when rotating the middle slice (layer) with the bottom layer, an extra twist restores the bottom layer. (The vertical slice is not used in solving the CUBE, but is used in pattern generations from a solid CUBE.)

' (prime): Sometimes at the beginning when doing the top 'f' face, I move the bottom layer and middle layer together.

ALIGN: An alignment step aligns the faces of the CUBE (a correction).

(xxx): A top 'f' layer cube that requires two operations since it is either on the bottom 'b' face or incorrectly in the top 'f' layer.

ADJUST: A step that places an edge cube, that is to be sent to the bottom 'b' layer, in a READY position.

THE FOLLOWING SEVEN EXERCISES WILL INCLUDE A SIMILAR TABLE AND A SUMMARY BUT THE NUMBER OF FIGURES IS REDUCED TO SEVEN AS FOLLOWS: (1) the scrambled CUBE; (2) the top done; (3) bottom corner cubes positioned; (4) the bottom corner cubes done; (5) the bottom edge cubes done, completing the bottom; (6) middle edge cubes positioned; and (7) CUBE is done. What is germane here besides the gist of solving the CUBE is the tabulation for Table 10EX. HOWEVER, A SEVEN-PAGE APPENDIX SHOWS THE MISSING COMPLEMENTARY FIGURES. (The addendum had been an after thought.)

CAUTION: Remember when doing R/V/1/1 translations on a particular face, that face is really determined by the four corner cubes and not the center cube since the middle layer is frequently being adjusted.

EXAMPLES

```
PSI(tv) .------------,          PS2(tv) .------------,
        / t / e / t /|                  / t / t / t /|
      /---/---/---/ r :-t,k,r         /---/---/---/ r :
      / f / t / k /|/|                / t / t / t /|/|
    /---/---/---/ b / e:            /---/---/---/ r / f:
   / t / b / k /|/|/|              / t / t / t /|/|/|
  :------------: e / r / b :      :------------: r / r / b :
  | e | e | b |/|/|/              | f | f | f |/|/|/
  :------------: f / t /          :------------: e / e /
  | t | f | t |/|/                | k | f | f |/|/
  :------------: b /              :------------: f /
  | t | e | r |/ FORTY bfe-|      | b | k | b |/-b,f,r
  '------------' TWISTS           '------------' top done

PSI(bv) .------------,           PS2(bv) .------------,
        / f / r / k /|                   / r / b / e /|
      /---/---/---/ r :                /---/---/---/ k :
      / f / b / f /|/|                 / b / b / f /|/|
    /---/---/---/ b / t:             /---/---/---/ r / b:
   / r / b / e /|/|/|               / e / r / r /|/|/|
  :------------: f / k / k :-tkr   :------------: k / k / k :
  | f | r | b|/|/|/                | f | k | b |/|/|/
  :------------: k / k /           :------------: r / k /
  | k | e | r |/|/                 | e | e | b |/|/
  :------------: e /               :------------: k /
  | k | r | f |/ FORTY             | e | e | e |/ top done
  '------------' TWISTS            '------------'

PS3(tv) .------------,            PS6(tv) .------------,
        / t / t / t /|                    / t / t / t /|
      /---/---/---/ r :                 /---/---/---/ r :
      / t / t / t /|/|                  / t / t / t /|/|
    /---/---/---/ r / f:              /---/---/---/ r / f:-b,f
   / t / t / t /|/|/|                / t / t / t /|/|/|
  :------------: r / r / b :        :------------: r / r / r :
  | f | f | f |/|/|/                | f | f | f |/|/|/
  :------------: e / k /            :------------: e / k /
  | k | f | f |/|/                  | k | f | f |/|/
  :------------: b /                :------------: r /
  | b | b | r |/ cor.               | f | f | f |/ corners
  '------------' pos. done          '------------' done

PS3(bv) .------------,            PS6(bv) .------------,
        / f / b / k /|                    / b / r / b /|
      /---/---/---/ r :                 /---/---/---/ k :
      / e / b / r /|/|                  / r / b / e /|/|
    /---/---/---/ f / b:              /---/---/---/ b / b :-b,f
   / e / r / e /|/|/|                / b / b / b /|/|/|
  :------------: b / k / k:         :------------: k / k / k :
  | f | k | k |/|/|/                | e | k | e |/|/|/
  :------------: r / k /            :------------: r / k /
  | e | e | b |/|/                  | e | e | b |/|/
  :------------: k /                :------------: k /
  | e | e | e |/ cor.               | e | e | e |/ corners
  '------------' pos. done          '------------' done
```

EXAMPLE #4 POSTSCRIPT!!!!

Figure PSI is a scrambled CUBE (after 40 twists with eyes closed) but the t,k,r top 't' corner cube is already okay. Figure PS2 is the CUBE after moving seven top layer cubes thereby finishing the top 't' layer. It took 26 steps including two alignments of the faces. Perusal of Figure PS2 shows that bottom corner cubes b,f,r and b,f,e are already positioned correctly in the bottom layer. Figure PS3 is the CUBE after doing a R/O operation on the front 'f' face that interchanges the other two bottom corners thereby positioning all four bottom corners correctly. Perusal of Figure PS3 shows that none of the bottom corners are oriented correctly. A random R/B/1/1 translation is done on one of the faces, the front face 'f is chosen. [PS4 and PS5 are not shown but are paraphrased as follows.] PS4 showed that after the R/B/1/1 random translation, bottom corner cubes b,k,r and b,k,e ended up being okay, an OOSI situation. PS5 showed the bottom corners 'in-sync' after a R/B/1/1 translation on the front 'f' face. Then two consecutive R/B/1/1 translations on the right 'r' face finished the bottom four corners as shown in Figure PS6 with corner cube b,f,r being the zero reference cube. Perusal of Figure PS6 shows that bottom edge cubes b,r and b,f can be put in a READY position by adjusting the middle layer two Q turns. [PS7 through PSI3 are not shown but are paraphrased as follows.] PS7 showed the b,r and b,f edge cubes in a READY position. Edge cube b,r was chosen and PS8 showed the CUBE after a R/V/1/1 translation on the front 'f' face sent the b,r edge cube to the bottom layer (bumped back the k,r edge cube). Aligning the faces (2Qt's were required) put the b,f edge cube in a READY position.

```
PS14(tv) .-----------,
        / t / t / t /|
       /---/---/---/ r :
      / t / t / t /|/|
     /---/---/---/ r / k :
    / t / t / t /|/|/|
    :-----------: r / r / r :
    | f | f | f |/|/|/|
    :-----------: r / r /
    | k | f | f |/|/
    :-----------: r /
    | f | f | f |/ bottom
    '-----------' done
PS14(bv) .-----------,
        / b / b / b /|
       /---/---/---/ k :
      / b / b / b /|/|
     /---/---/---/ k / e :
    / b / b / b /|/|/|
    :-----------: k / k / k :
    | e | e | e |/|/|/|
    :-----------: e / k /
    | r | e | f |/|/
    :-----------: k /
    | e | e | e |/ bottom
    '-----------' done
PS15(tv) .-----------,
        / t / t / t /|
       /---/---/---/ r :
      / t / t / t /|/|
     /---/---/---/ r / r :
    / t / t / t /|/|/|
    :-----------: r / r / r :
    | f | f | f |/|/|/|
    :-----------: r / r /
    | e | f | f |/|/
    :-----------: r /
    | f | f | f |/ middle
    '-----------' pos. done
PS15(bv) .-----------,
        / b / b / b /|
       /---/---/---/ k :
      / b / b / b /|/|
     /---/---/---/ k / k :
    / b / b / b /|/|/|
    :-----------: k / k / k :
    | e | e | e |/|/|/|
    :-----------: e / k /
    | f | e | k |/|/
    :-----------: k /
    | e | e | e |/ middle
    '-----------' pos. done
```

PS9 showed the CUBE after a R/V/1/1 translation on the left 'e' face sent the b,f edge cube to the bottom layer (bumped back the f,r edge cube). 'TWO MORE TO GO.' PS9 showed along with Figure PS6 shows that edge cubes b,e and b,k are wrongly located in the bottom layer, thus cannot be put in a READY position. Therefore, one of them needs to be bumped back and b,e was chosen. PS10 showed the CUBE after a R/V/1/1 translation on the right 'r' face bumped back the b,e edge cube. PS11 then showed the b,e edge cube in a READY position, however two adjustments were first required. First the middle layer needed to be adjusted two Q turns and then the back 'k' face was inverted. PS12 showed the CUBE after a R/V/1/1 translation on the back 'k' face sent the b,e edge cube to the bottom layer (bumped back the b,k edge cube) and the back 'k' face re-inverted. Adjusting the middle layer one Q turn to the right put the b,k edge cube in a READY position. PS13 showed the CUBE after a R/V/1/1 translation on the right 'r' face sent the b,k edge cube to the bottom layer (bumped back the f,r edge cube). Figure PS14 shows the CUBE after the faces are aligned by a one Q turn of the middle layer to the right which finishes the bottom layer. Perusal of Figure PS14 shows that only one of the four middle edge cubes is position-ed correctly and that is the f,r edge cube. Doing two consecutive RM/0 operations on the right 'r' face correctly positions the four middle edge cubes as shown in Figure PS15. Perusal of Figure PS15 shows that the CUBE is done except for orienting the two middle edge cubes k,e and f,e via an ALG01 operation. 'SO' doing a R/V/1/1 translation on the left 'e' face sends edge cube f,e to the

```
PS16(tv) .-----------,
        / t / t / t /|
       /---/---/---/ r :
      / t / t / t /|/|
     /---/---/---/ r / r :
    / t / t / t /|/|/|
    :-----------: r / r / r :
    | f | f | f |/|/|/|
    :-----------: r / r /
    | f | f | f |/|/
    :-----------: r /
    | f | f | f |/ CUBE
    '-----------' DONE
PS16(bv) .-----------,
        / b / b / b /|
       /---/---/---/ k :
      / b / b / b /|/|
     /---/---/---/ k / k :
    / b / b / b /|/|/|
    :-----------: k / k / k :
    | e | e | e |/|/|/|
    :-----------: k / k /
    | e | e | e |/|/
    :-----------: k /
    | e | e | e |/ CUBE
    '-----------' DONE
```

to the bottom layer, next a R/V/1/1 translation on the front 'f' face sends edge cube k,e to the bottom layer, then a C/B/1/2 translation on the back 'k' face inverts them. Now, a middle layer one Q turn adjustment to the left enables a R/V/1/1 translation on the left 'e' face to send the f,b edge cube (only one in a READY position) to the bottom layer, then a middle layer two Q turn adjustment enables a R/V/1/1 translation on the front 'f' face to send the r,b edge cube (only one in a READY position) to the bottom layer, finally, a R/V/1/1 translation on the back 'k' face (no adjustment) sends the e,b edge cube (only one in a READY position) to the bottom layer finishing ALG01 after the faces are aligned. The CUBE is done. This forth exercise took 162 steps (213 twists). SO KEEP ON KEEPING ON TWIST TWISTING YOUR CUBE!!!!

EXAMPLES

EXERCISE-FOUR SUMMARY (see Table PS)

a) Seven of the eight top layer cubes are okay after 26 steps moved them from the side of either the middle layer or the bottom layer which included two alignments. One corner cube in the top layer was already okay.

b) None of the top 't' layer cubes ended up incorrectly on the top layer nor were corner cubes facing the bottom, therefore, no bumping was needed.

c) Somewhat similar to Exercise I, the bottom corners are positioned correctly after a R/O operation interchanges two adjacent corner cubes. (However, for Exercise I the two corner cubes to be interchanged were on a diagonal.)

d) Unlike the other three exercises, none of the bottom corner cubes are oriented correctly (are okay). A random R/B/I/I translation on either of the four faces f, r, k, or e is required. (Table 10EX at the end of this 'exercise' section shows that the necessity for such a random translation on one of the faces occurs four times in these ten exercises.) An OOSI routine uses a R/B/I/I translation to get the bottom corners 'in-sync' so that two subsequent R/B/I/I translations finishes the bottom four corner cubes.

e) The four bottom edge cubes are okay after four R/V/I/I translations move them to the bottom layer after they are 'adjusted' to a READY position. Unlike the other three exercises, after making two-bottom edge cubes okay the other two were already in the bottom layer (incorrectly) so adjusting the middle layer couldn't put them in a READY position. So an extra R/V/I/I translation was used to bump one of them back so it could be put in a READY position. However, similar to the other three exercises, before the R/V/I/I translation a face needed to be inverted (2Qt's) in order to put one of these edge cubes in a READY position. (Table 10EX at the end of this 'exercise' section shows that the necessity for such a bumping-back operation occurs thrice in these ten exercises.)

f) Like Exercise I the four middle edge cubes are correctly positioned after two subsequent RM/O operations rotates twice three of the four middle edge cubes. Two of them are oriented correctly (are okay).

g) The CUBE is done after an ALGO1 operation orients the two remaining adjacent middle edge cubes.

NOTES:

TABLE PS   EXERCISE-FOUR

OPER-ATION	STEPS	TWISTS		(Bottom Corners Cont'd)			(DO MIDDLE EDGES)		
				R/B/1/1	8	--	RM/0	4	2
				R/B/1/1	8	--	RM/0	4	2
t,k,r 't' cube	OK	--		R/B/1/1	8	--			
t,k 't' cube	3'	--		R/B/1/1	8	--	(ALGO1)		
Align	1'	--					R/V/1/1	7	3
t,f,e 't' cube	4	--		Corners Done	41	41	R/V/1/1	7	3
t,k,e 't' cube	3	--					C/B/1/2	7	4
t,f,r 't' cube	4	--		(DO BOTTOM EDGES)			Adjust(-1Qt)	1	1
t,e 't' cube	3	2		Adjust	1	1	R/V/1/1	7	3
t,f, 't' cube	3	2		R/V/1/1	7	3	Adjust(+2Qt)	1	1
t,r 't' cube	4	1		Align	1	1	R/V/1/1	7	3
Align	1'	--		R/V/1/1	7	3	R/V/1/1	7	3
				R/V/1/1(bump)	7	3	Align	1	1
Top Done	26	31		Adjust	1	1	(ALGO1 Done	45	67)
				Invert 'k'	1	--			
(DO BOTTOM CORNERS)				R/V/1/1	7	3	Edges Done	53	79
R/0	9	--		Re-invert 'k'	1	--			
				Adjust	1	1	CUBE DONE	162	213
				R/V/1/1	7	3			
				Align	1	1			
				Bottom done	42	62			

TABLE PS   NOTES

STEPS: A step is one twist of the CUBE that can be one 'Q' turn or two 'Q' turns, e.g. inverting a face is two 'Q' turns.

TWISTS: A twist column shows when a 'step' is two twists instead of one twist, e.g. when bringing down the center slice along with the right side, an extra twist restores the right side; similarly, when rotating the middle slice (layer) with the bottom layer, an extra twist restores the bottom layer. (The vertical slice is not used in solving the CUBE, but is used in pattern generations from a solid CUBE.)

' (prime): Sometimes at the beginning when doing the top 't' face, I move the bottom layer and middle layer together.

ALIGN: An alignment step aligns the faces of the CUBE (a correction).

(xxx): A top 't' layer cube that requires two operations since it is either on the bottom 'b' face or incorrectly in the top 't' layer.

ADJUST: A step that places an edge cube, that is to be sent to the bottom 'b' layer, in a READY position.

```
J1(tv)    .----------,          J3(tv)    ,-----------,
       / t / t / e /|                  / t / t / t /|
      /---/---/---/ K:                /---/---/---/ r:
     / b / t / r /|/|                / t / t / t /|/|
    /---/---/---/ f/ f:             /---/---/---/ r/ f:
   / e / b / b /|/|/|              / t / t / t /|/|/|
   :----------: f/ r/ e:           :----------: r/ r/ b:
   | K | K | r |/|/|/|             | f | f | f |/|/|/|
   :----------: t/ K/              :----------: K/ r/
   | f | f | e |/|/|               | K | f | r |/|/|
   :----------: f/                 :----------: f/
   | K | K | t |/TWENTY-      #-| K | e | e |/-b,e,f
   '----------'FIVE TWISTS         '----------' top done

J1(bv)    .----------,          J3(bv)    ,-----------,
       / e / r / b /|                  / b / b / K /|
      /---/---/---/ f:    #=b,e,K     /---/---/---/ r:
     / t / b / b /|/|                / b / b / K /|/|
    /---/---/---/ r/ e:             /---/---/---/ e/ e:
   / b / e / r /|/|/|              / e / f / r /|/|/|
   :----------: f/ K/ b:           #-:----------: b/ K/ K:
   | r | K | t |/|/|/|             | b | r | f |/|/|/|
   :----------: b/ r/              :----------: f/ K/
   | t | e | f |/|/|               | b | e | b |/|/|
   :----------: r/                 :----------: K/
   | t | e | K |/ TWENTY-          | e | e | e |/ top
   '----------'FIVE TWISTS         '----------' done

J5(tv)    .----------,          J6(tv)    ,-----------,
       / t / t / t /|                  / t / t / t /|
      /---/---/---/ r:                /---/---/---/ r:
     / t / t / t /|/|                / t / t / t /|/|
    /---/---/---/ r/ f:             /---/---/---/ r/ f:
   / t / t / t /|/|/|              / t / t / t /|/|/|
   :----------: r/ r/ b:           :----------: r/ r/ r:
   | f | f | f |/|/|/|             | f | f | f |/|/|/|
   :----------: K/ K/              :----------: K/ r/
   | K | f | r |/|/|               | K | f | r |/|/|
   :----------: r/                 :----------: r/
   | e | e | f |/ cor. pos.        | f | e | f |/ corners
   '----------' done               '----------' done

J5(bv)    .----------,          J6(bv)    ,-----------,
   f,r,b-/ b / e / K /|                  / b / f / b /|
      /---/---/---/ r:                /---/---/---/ K:
     / b / b / f /|/|                / b / b / r /|/|
    /---/---/---/ r/ e:             /---/---/---/ b/ e:
   / f / r / e /|/|/|              / b / e / b /|/|/|
   :----------: b/ K/ K:           :----------: K/ K/ K:
   | b | b | K |/|/|/|             | e | K | e |/|/|/|
   :----------: f/ K/              :----------: f/ K/
   | b | e | b |/|/|               | b | e | b |/|/|
   :----------: K/                 :----------: K/
   | e | e | e |/ cor. pos.        | e | e | e |/ corners
   '----------' done               '----------' done
```

EXAMPLE #5 JUST-IN CASE!!!!!

Figure J1 is a scrambled CUBE (after 25 twists with eyes closed). Figure J3 is the CUBE after moving six top 't' face cubes via the FOUR STEP APPROACH thereby finishing the top layer. It took 30 steps including two alignments and two dumps (moving-out incorrect top 't' face cubes from the top layer). FIG. J2 is not shown here which showed the CUBE after the top face was aligned one Q turn to the left making the top layer cubes t,r and t,r,K okay. Perusal of Figure J3 shows that after rotating the bottom layer one Q turn to the left, bottom corner cubes b,e,f and b,e,K are positioned correctly. Figure J5 shows the CUBE after doing a R/O operation on the left 'e' face that interchanges the other two bottom corners thereby positioning correctly all four bottom corners. FIG. J4 is not shown here which showed the corner cubes b,e,f and b,e,K correctly positioned after the above rotation. Perusal of Figure J5 shows that bottom corner cube f,r,b is okay and Lo and Behold the corners are already 'in-sync'. Figure J6 shows the CUBE after a R/B/1/1 translation on the right 'r' face finishes the bottom layer. Perusal of Figure J6 shows that edge cube f,b can be put in a READY position to bump back the b,e edge cube by adjusting the middle layer one Q turn to the right. [FIG'S. J7 through J12 are not shown but are paraphrased as follows.] Fig. J7 showed the CUBE after edge cube f,b is in a READY position and a R/V/1/1 translation on the left 'e' face sent it to the bottom layer. Similar to f,b above, a 1Qt adjustment of the middle layer to the right put the e,b edge cube in a READY position to bump back edge cube e,K. FIG. J8 showed the CUBE after edge cube e,b is in a READY position and a R/V/1/1 translation on the

```
J13(tv)  .-----------,
        / t / t / t / |
       /---/---/---/ r :
      / t / t / t / | /|
     /---/---/---/ r / r :
    / t / t / t / |/|/|
   :-----------: r/ r/ r :
   | f | f | f | |/|/|/|
   :-----------: e/ r /
   | k | f | f | |/|/
   :-----------: r /
   | f | f | f | / bottom
   '-----------' done
J13(bv)  .-----------,
        / b / b / b / |
       /---/---/---/ k :
      / b / b / b / | /|
     /---/---/---/ k / f |
    / b / b / b / |/|/|
   :-----------: k/ k/ k :
   | e | e | e | |/|/|/|
   :-----------: k/ k /
   | r | e | e | |/|/
   :-----------: k /
   | e | e | e | / bottom
   '-----------' done
J14(tv)  .-----------,
        / t / t / t / |
       /---/---/---/ r :
      / t / t / t / | /|
     /---/---/---/ r / r :
    / t / t / t / |/|/|
   :-----------: r/ r/ r :
   | f | f | f | |/|/|/|
   :-----------: r/ r /
   | f | f | f | |/|/
   :-----------: r / middle pos.
   | f | f | f | /   done,
   '-----------' (CUBE DONE)
J14(bv)  .-----------,
        / b / b / b / |
       /---/---/---/ k :
      / b / b / b / | /|
     /---/---/---/ k / k :
    / b / b / b / |/|/|
   :-----------: k/ k/ k :
   | e | e | e | |/|/|/|
   :-----------: k/ k /
   | e | e | e | |/|/
   :-----------: k / middle pos.
   | e | e | e |/    done,
   '-----------' (CUBE DONE)
```

back 'k' face sent it to the bottom layer. FIG. J8 showed that a double adjustment was needed to put edge cube b,k in a READY position to bump back edge cube b,r. First the middle layer was adjusted one Q turn to the left. Secind, the right 'r' face was inverted. FIG. J9 showed edge cube k,b in a READY position. FIG. J10 showed the CUBE after edge cube k,b is in a READY position and a R/V/I/I translation on the right face sent it to the bottom layer. FIG. J10 also showed that after re-inverting the right 'r' face, inverting the front 'f face put the edge cube r,b in a READY position to bump back the f,r edge cube. FIG. J11 showed edge cube r,b in a READY position. FIG. J12 showed the CUBE after edge cube r,b is in a READY position and a R/V/I/I translation on the front 'f face sent it to the bottom completing the bottom layer. After re-inverting the front 'f face and aligning the CUBE (middle layer one Q turn to the left), Figure J13 shows the CUBE with the bottom layer completed. Perusal of Figure J13 shows that middle edge cube e,k is okay but is the only middle edge cube positioned correctly. So one RM/O operation on the left 'e' face rotates the other three middle edge cubes and Lo and Behold the other three happened to turn up correctly positioned and oriented finishing the CUBE as shown in Figure J14. Exercise five took only 88 steps (110 twists).

EXERCISE-FIVE SUMMARY (see Table J)

a) Four of the eight top face cubes are okay after 15 steps move them from the side of either the middle or bottom layer. One corner cube and one edge cube in the top layer were already okay.

b) The other two top face cubes are okay after fourteen steps for moving-out and then moving two corner cubes whose 't' face was on the bottom layer.

c) Same as for Exercises 2 and 3, since at least two bottom corners must be correctly positioned, the bottom layer is further rotated until two correctly positioned corners are found. Then all four-bottom corner cubes are positioned correctly after a R/O operation interchanges two adjacent corners.

d) Then again like Exercises 2 and 3, by happenstance one bottom corner cube is okay and the other three corners are 'in-sync'. However, unlike Exercise 2 but like Exercise 3 only one instead of two R/B/I/I translations finishes the bottom four corner cubes (they are all okay).

e) The four bottom edge cubes are okay after four R/V/I/I translations move them to the bottom layer after they are 'adjusted' to a READY position. However, like Exercises 1, 2, and 3, before the R/V/I/I translation a face needed to be inverted (2Qt's) in order to put two of these edge cubes in a READY position.

f) The four middle edge cubes are correctly positioned after only one RM/O operation which rotates three of the four middle edge cubes. Similar to Exercise 3, all four edge cubes instead of two are correctly oriented after a RM/O operation. An ALGOl operation, therefore is not needed since the CUBE is finished.
   (Table 10EX at the end of this exercise section shows that this immediate orientation of these four middle edge cubes occurs twice in these ten exercises. Here and in Exercise 3.)

TABLE J   EXERCISE-FIVE

OPER-ATION	STEPS	TWISTS		(DO BOTTOM CORNERS)				(bottom edges cont'd)		
				Align	1	--		R/V/1/1	7	3
				R/0	9	--		Re-invert 'f'	1	--
Align	1	--		R/B/1/1	8	--		Align	1	1
t,k,r 't' cube	OK'	--		-----------------				-----------------		
t,r 't' cube	OK	--		Corners Done	18	18		Bottom Done	36	52
t,k 't' cube	4	2		================				================		
t,e 't' cube	3'	--								
Align	1'	--		(DO BOTTOM EDGES)				(DO MIDDLE EDGES)		
t,f 't' cube	3	2		Adjust(1Qt)	1	1		RM/01	4	2
t,r,f 't' cube	4	--		R/V/1/1	7	3		-----------------		
(t,e,f) 't' cube	3	--		Adjust(1Qt)	1	1		Edges Done	4	6
t,e,f 't' cube	4	--		R/V/1/1	7	3		================		
(t,e,k) 't' cube	3	--		Adjust(1Qt)	1	1		CUBE DONE	88	110
t,e,k 't' cube	4	--		Invert 'r'	1	--		================		
-----------------				R/V/1/1	7	3				
Top Done	30	34		Re-invert 'r'	1	--				
================				Invert 'f'	1	--				

TABLE J   NOTES

STEPS: A step is one twist of the CUBE that can be one 'Q' turn or two 'Q' turns, e.g. inverting a face is two 'Q' turns.

TWISTS: A twist column shows when a 'step' is two twists instead of one twist, e.g. when bringing down the center slice along with the right side, an extra twist restores the right side; similarly, when rotating the middle slice (layer) with the bottom layer, an extra twist restores the bottom layer. (The vertical slice is not used in solving the CUBE, but is used in pattern generations from a solid CUBE.)

' (prime):   Sometimes at the beginning when doing the top 't' face, I move the bottom layer and middle layer together.

ALIGN: An alignment step aligns the faces of the CUBE (a correction).

(xxx):   A top 't' layer cube that requires two operations since it is either on the bottom 'b' face or incorrectly in the top 't' layer.

ADJUST: A step that places an edge cube, that is to be sent to the bottom 'b' layer, in a READY position.

NOTES:

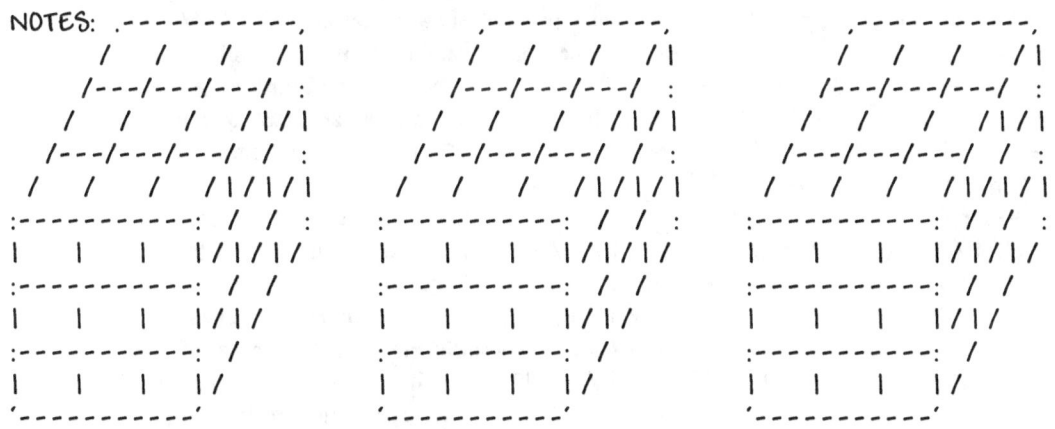

```
W1(tv)   .----------,
        / k / r / e /|
       /---/---/---/ b:
      / r / t / e /|/|
     /---/---/---/ b/ e:
    / f / k / t /|/|/|
    :----------:k/ r/ r:
    | e | r | r |/|/|/
    :----------: e/ K/
    | b | f | K |/|/
    :----------: K/
    | t | t | r |/  FIFTEEN
    '----------'    TWISTS

W1(bv)   .----------,
        / b / t / f /|
       /---/---/---/ b:
      / f / b / e /|/|
     /---/---/---/ f/ t:
    / r / b / e /|/|/|
    :----------: K/ K/ f:
    | f | K | t |/|/|/
    :----------: r/ t/
    | f | e | f |/|/
    :----------: b/
    | t | b | e |/  FIFTEEN
    '----------'    TWISTS

W5(tv)   .----------,
        / t / t / t /|
       /---/---/---/ r:
      / t / t / t /|/|
     /---/---/---/ r/ b:
    / t / t / t /|/|/|
    :----------: r/ r/ b:
    | f | f | f |/|/|/
    :----------: b/ e/
    | e | f | f |/|/
    :----------: f/
    | b | r | b |/  cor. pos.
    '----------'    done

W5(bv)   .----------,
        / r / f / K /|
       /---/---/---/ r:
      / K / b / b /|/|
     /---/---/---/ r/ K:
    / e / e / K /|/|/|
    :----------: e/ K/ K:
    | f | b | b |/|/|/
    :----------: f/ K/
    | K | e | r |/|/
    :----------: K/
    | e | e | e |/  cor. pos.
    '----------'    done
```

```
W3(tv)   ,----------,
        / t / t / t /|
       /---/---/---/ r:
      / t / t / t /|/|
     /---/---/---/ r/ b:
    / t / t / t /|/|/|
    :----------: r/ r/ f:
    | f | f | f |/|/|/
    :----------: b/ e/
    | e | f | f |/|/
    :----------: b/
    | b | b | f |/  top
    '----------'    done

W3(bv)   ,----------,
        / e / f / b /|
       /---/---/---/ r:
      / e / b / r /|/|
     /---/---/---/ K/ K:
    / r / r / K /|/|/|
    :----------: e/ K/ K:
    | K | b | b |/|/|/
    :----------: f/ K/
    | K | e | r |/|/
    :----------: K/
    | e | e | e |/  top
    '----------'    done

W8(tv)   ,----------,
        / t / t / t /|
       /---/---/---/ r:
      / t / t / t /|/|
     /---/---/---/ r/ b:
    / t / t / t /|/|/|
    :----------: r/ r/ b:
    | f | f | f |/|/|/
    :----------: b/ r/-r,b
    | e | f | f |/|/
    :----------: r/
    | f | b | f |/  corners
    '----------'    done

W8(bv)   ,----------,
        / b / b / b /|
       /---/---/---/ K:
  e,b-/ e / b / f /|/|
     /---/---/---/ e/ K:
    / b / K / b /|/|/|
    :----------: K/ K/ K:
    | e | r | e |/|/|/
    :----------: f/ K/
    | K | e | r |/|/
    :----------: K/
    | e | e | e |/  corners
    '----------'    done
```

EXAMPLE #6 <u>WOW !!!!!!</u>

Figure W1 is a scrambled CUBE (after 15 twists with eyes closed). Figure W3 is the CUBE after moving seven top 't' face cubes via the FOUR STEP APPROACH finishing the top layer. It took 33 steps including two alignments and two dumps (removing two incorrect top cubes from the bottom layer). FIG. W2 is not shown here which showed the CUBE after the top face was aligned one Q turn to the right making top layer cube t,r,k okay. Perusal of Figure W3 shows that only the bottom corner cube b,e,k is positioned correctly. Consequently, rotating the bottom layer one Q turn to the left positions two bottom corner cubes correctly, the b,e,f and b,r,f corner cubes. Figure W5 shows the CUBE after doing an R/0 operation on the front 'f face that interchanges the other two bottom corner cubes thereby positioning correctly all four bottom corners. FIG. W4 is not shown here which showed the corner cubes b,e,f and b,r,f correctly positioned after the above rotation. But now Figure W5 shows that not one of the bottom four corner cubes is correctly oriented, so an R/B/l/l translation is to be done on one of the four faces, the front 'f face is chosen. [W6 and W7 are not shown but are paraphrased as follows.] W6 showed that after the R/B/l/l random translation, bottom corner cubes b,r,f and b,r,k ended up being okay, an OOS2 situation. W7 showed the bottom corners 'in-sync' after a R/B/l/l translation on the front 'f face. Then two consecutive R/B/l/l translations on the left 'e' face finished the bottom four corner cubes as shown in Figure W8. Perusal of Figure W8 shows that by rotating the middle layer one Q turn to the left puts the bottom edge cube f,b in a READY position to bump back

```
W12(tv)  .----------,
        / t / t / t / |
       /---/---/---/ r :
      / t / t / t / | / |
     /---/---/---/ r / f :
    / t / t / t / | / | / |
   :----------: r / r / r :
   | f | f | f | / | / | /
   :----------: k / r /
   | r | f | r | / | /
   :----------: r /
   | f | f | f | / bottom
   '----------'  done
W12(bv)  .----------,
        / b / b / b / |
       /---/---/---/ k :
      / b / b / b / | / |
     /---/---/---/ k / e :
    / b / b / b / | / | / |
   :----------: k / k / k :
   | e | e | e | / | / | /
   :----------: e / k /
   | f | e | k | / | /
   :----------: k /
   | e | e | e | / bottom
   '----------'  done
W13(tv)  .----------,
        / t / t / t / |
       /---/---/---/ r :
      / t / t / t / | / |
     /---/---/---/ r / k :
    / t / t / t / | / | / |
   :----------: r / r / r :
   | f | f | f | / | / | /
   :----------: f / r /
   | e | f | r | / | /
   :----------: r /
   | f | f | f | / middle
   '----------'  pos. done
W13(bv)  .----------,
        / b / b / b / |
       /---/---/---/ k :
      / b / b / b / | / |
     /---/---/---/ k / r :
    / b / b / b / | / | / |
   :----------: k / k / k :
   | e | e | e | / | / | /
   :----------: e / k /
   | f | e | k | / | /
   :----------: k /
   | e | e | e | / middle
   '----------'  pos. done
```

edge cube e,b. Also, Figure W8 shows that bottom edge cube r,b is already done, a FREEBY. [W9 through W11 are not shown but are paraphrased as follows.] W9 showed the CUBE after rotating the middle layer and then doing an R/V/1/1 translation on the left 'e' face that sent the f,b cube to the bottom layer. Then W10 showed that by inverting the back 'k' face put b,e edge cube in a READY position to bump back edge cube r,k. W11 showed the CUBE after a R/V/1/1 translation on the back 'k' face sent the b,e edge cube to the bottom layer and also showed that edge cube k,b was already in a READY position to bump back edge cube e,f, however in order to have done this a R/V/1/1 translation was done on the left 'e' face but with the CUBE upside down. Figure W12 shows the CUBE after the k,b edge cube was sent to the bottom layer, after the back 'k' face was re-inverted, and after the middle layer was realigned one Q turn to the right. NOTE, for W9, W10, and W11 the center 'r' cube appeared on the front 'f face, the 'k' center cube on the right 'r' face, etc. Figure W12 shows that the bottom layer is done. Perusal of Figure W12 shows that middle edge cube e,k is already positioned correctly but not oriented correctly. Figure W13 shows that after doing two RM/O operations on the left 'e' face that rotates the other three middle edge cubes, all four middle edge cubes are positioned correctly but none of them are oriented correctly, a worst case scenario. Consequently, the CUBE is done after doing an ALGO1 operation on the front 'f face followed by another ALGO1 operation on the back 'k' face but it takes 98 steps (146 twists) just for these two ALGO1 operations. The total steps are 198 (265 twists).

```
W14(tv)  .----------,
        / t / t / t / |
       /---/---/---/ r :
      / t / t / t / | / |
     /---/---/---/ r / r :
    / t / t / t / | / | / |
   :----------: r / r / r :
   | f | f | f | / | / | /
   :----------: r / r /
   | f | f | f | / | /
   :----------: r /
   | f | f | f | / CUBE
   '----------'  DONE
W14(bv)  .----------,
        / b / b / b / |
       /---/---/---/ k :
      / b / b / b / | / |
     /---/---/---/ k / k :
    / b / b / b / | / | / |
   :----------: k / k / k :
   | e | e | e | / | / | /
   :----------: k / k /
   | e | e | e | / | /
   :----------: k /
   | e | e | e | / CUBE
   '----------'  DONE
```

EXERCISE-SIX SUMMARY
a) Five of the eight top layer cubes are okay after 19 steps moves them from the side of either the middle or top layer which included one alignment. One top corner cube was okay after an alignment of the top layer.
b) The remaining two top 't' face corner cubes took 13 steps for moving-out and then moving them to the top layer since their top 't' face was on the bottom layer.
c) Same as for Exercises 2, 3, and 5, since at least two bottom corners must be positioned correctly, the bottom layer is further rotated until two correctly positioned corner cubes are found. Then all four-bottom corner cubes are positioned correctly after a R/O operation interchanges two adjacent corners.
d) Like Exercise 4 none of the bottom corner cubes are oriented correctly (are okay). A random R/B/1/1

translation on either of the four faces f, r, k, or e is required. (Table 10EX at the end of this 'exercise' section shows that the necessity for such a random translation on one of the faces occurs four times in these ten exercises.) An OOS2 routine uses a R/B/1/1 translation to get the bottom corners 'in-sync' so that two subsequent R/B/1/1 translations finishes the bottom four corner cubes.

e) The four bottom edge cubes are okay after three R/V/1/1 translations move three of them to the bottom layer after they are in a READY position. One of the three bottom edge cubes is already okay. (Table 10EX at the end of this 'exercise' section shows that on three occasions one of the four bottom edge cubes is already okay.) However, before the R/V/1/1 translation a face needed to be inverted (2Qt's) then two of the three-bottom edge cubes are put in a READY position.

f) Like Exercise 1 the four middle edge cubes are correctly positioned after two subsequent RM/0 operations rotates twice three of the four middle edge cubes. However, unlike any of the previous exercises, none of the four middle edge cubes are oriented correctly (are okay).

g) The CUBE is done after two subsequent ALG01 operations orient all four middle edge cubes. (Table 10EX at the end of this 'exercise' section shows that on just two occurrences this worst case happens when none of the middle edge cubes are oriented correctly.)

TABLE W   EXERCISE-SIX

OPER-ATION	STEPS	TWISTS		(DO BOTTOM CORNERS)				(bottom edges cont'd)		
				Align	1	--		One cube OK --		--
				R/0	9	--				
Align	1	--		R/B/1/1	8	--		Bottom Done 25		36
t,k,r 't' cube	OK	--		R/B/1/1	8	--				
t,k 't' cube	4	2		R/B/1/1	8	--		(DO MIDDLE EDGES)		
t,f 't' cube	3	2		R/B/1/1	8	--		RM/0	4	2
t,r,f 't' cube	4	--						RM/0	4	2
t,e 't' cube	4	2		Corners Done 42		42		ALG01	45	22
Align	1	--						ALG01	45	22
t,r 't' cube	3	2		(DO BOTTOM EDGES)						
(t,e,k) 't' cube	3	--		Adjust(1Qt)	1	1		Edges Done 98		146
t,e,k 't' cube	3	--		R/V/1/1	7	3				
(t,e,f) 't' cube	4	--		Invert 'k'	1	--		CUBE DONE 198		265
t,e,f 't' cube	3	--		R/V/1/1	7	3				
				R/V/1/1	7	3				
Top Done	33	41		Re-invert 'k'	1	--				
				Align	1	1				

TABLE W   NOTES

STEPS: A step is one twist of the CUBE that can be one 'Q' turn or two 'Q' turns, e.g. inverting a face is two 'Q' turns.

TWISTS: A twist column shows when a 'step' is two twists instead of one twist, e.g. when bringing down the center slice along with the right side, an extra twist restores the right side; similarly, when rotating the middle slice (layer) with the bottom layer, an extra twist restores the bottom layer. (The vertical slice is not used in solving the CUBE, but is used in pattern generations from a solid CUBE.)

' (prime): Sometimes at the beginning when doing the top 't' face, I move the bottom layer and middle layer together.

ALIGN: An alignment step aligns the faces of the CUBE (a correction).

(xxx): A top 't' layer cube that requires two operations since it is either on the bottom 'b' face or incorrectly in the top 't' layer.

ADJUST: A step that places an edge cube, that is to be sent to the bottom 'b' layer, in a READY position.

```
X1(tv)   .-----------,        X2(tv)   ,-----------,              EXAMPLE #7
       / f / k / t / l-t,k,e         / t / t / t / l       Figure X1 is a scrambled CUBE
      /---/---/---/ k :            /---/---/---/ r :        (after 25 twists with eyes closed).
     / b / t / k / l / l         / t / t / t / l / l       Figure X2 is the CUBE after
    /---/---/---/ b / t:        /---/---/---/ r / f:        moving six top 't' face cubes
   / k / b / b / l / l / l     / t / t / t / l / l / l      that were on the sides of either
  :-----------: e / r / k:    :-----------: r / r / k:-b,r,k the middle layer or the bot-
  l e l e l f l / l / l /     l f l f l f l / l / l /       tom layer. Corner cube t,k,e is
  :-----------: r / e /       :-----------: k / b /         aligned by moving the top layer 't' one Q
  l r l f l t l / l /         l e l f l e l / l /           turn to the right. Also aligning edge cube
  :-----------: e /           :-----------: b /             t,r with corner cube t,r,f before transfer
  l r l e l f l / TWENTY-     l f l b l f l/ top            to the top layer gave us the t,r edge
  '-----------'FIVE TWISTS    '-----------' done            cube as a freebie. It took 24 steps in-
X1(bv)   .-----------,        X2(bv)   ,-----------,        cluding three alignments and one
       / t / t / r / l              / e / f / r / l-b,r,k   bump (removing incorrect
      /---/---/---/ b :            /---/---/---/ b :        cube from the top layer) to finish
     / f / b / k / l / l         / r / b / k / l / l        the top 't' layer. After rotating
    /---/---/---/ e / f:        /---/---/---/ b / e:        the bottom layer (an alignment),
   / f / f / k / l / l / l     / r / r / e / l / l / l      Figure X2 shows that the bottom
  :-----------: r / k / e :-tke :-----------: b / k / k:    corner cubes b,r,k and b,e,k
  l t l b l t l / l / l /      l b l f l k l / l / l /      are positioned correctly so that
  :-----------: r / t /        :-----------: r / k /        a R/O operation on the back 'k' face
  l f l e l k l / l /          l b l e l k l / l /          positions correctly all four bottom corner
  :-----------: b /            :-----------: k /            cubes as shown in Figure X3. Perusal of
  l b l r l r l / TWENTY-      l e l e l e l / top          Figure X3 shows that not one of the
  '-----------'FIVE TWISTS     '-----------' done           bottom four corner cubes is okay
X3(tv)   .-----------,        X5(tv)   ,-----------,        (correctly oriented), so an R/B/1/1
       / t / t / t / l              / t / t / t / l         translation is to be done on one
      /---/---/---/ r :            /---/---/---/ r :        of the four faces, the front 'f' face
     / t / t / t / l / l         / t / t / t / l / l        is chosen. [FIG. X4 is not shown
    /---/---/---/ r / f:        /---/---/---/ r / f:        here but is paraphrased as fol-
   / t / t / t / l / l / l     / t / t / t / l / l / l      lows.] FIG. X4 showed that after
  :-----------: r / r / k:    :-----------: r / r / r:      the random translation on the
  l f l f l f l / l / l /     l f l f l f l / l / l /       front 'f' face, bottom corner cube
  :-----------: k / b /       :-----------: k / r /-r,b OK  f,r,b is okay and Lo-and-
  l e l f l e l / l /         l e l f l e l / l /           Behold the other three bottom
  :-----------: f /           :-----------: r /             corner cubes are already 'in-
  l e l r l b l / cor. pos.   l f l b l f l / corners       sync'. Thereby, Figure X5 shows
  '-----------' done          '-----------' done            the CUBE after one R/B/1/1
X3(bv)   .-----------,        X5(bv)   ,-----------,        translation on the right 'r' face
       / r / f / r / l              / b / b / b / l         orients the other three corner
      /---/---/---/ b :            /---/---/---/ k :        cubes finishing the bottom four
     / b / b / f / l / l         / f / b / k / l / l        corner cubes. Perusal of Figure
    /---/---/---/ r / e:        /---/---/---/ b / e:        X5 shows that bottom edge cube
   / f / k / k / l / l / l     / b / f / b / l / l / l      r,b is already okay, a freebie.
  :-----------: e / k / k:    :-----------: k / k / k:      Also, Figure X5 shows that by
  l b l b l b l / l / l /     l e l r l e l / l / l /       moving the middle layer two Q turns
  :-----------: r / k /       :-----------: r / k /         and then inverting the back 'k' face
  l b l e l k l / l /         l b l e l k l / l /           puts the e,b edge cube in a READY posi-
  :-----------: k /           :-----------: k /             tion to bump back edge cube f,r. [FIG.
  l e l e l e l / cor. pos.   l e l e l e l / corners       X6 through FIG. X9 are not shown
  '-----------' done          '-----------' done            but are paraphrased as follows.]
```

```
X10(tv)  .-----------,
        / t / t / t / |
       /---/---/---/ r :
       / t / t / t / | /1
      /---/---/---/ r / e :
      / t / t / t / | / | /1
     :-----------: r / r / r :
     | f | f | f | / | / | /1
     :-----------: f / r /
     | e | f | r | / | /
     :-----------: r /
     | f | f | f | / bottom
     '-----------' done
X10(bv)  .-----------,
        / b / b / b / |
       /---/---/---/ k :
       / b / b / b / | /1
      /---/---/---/ k / f :
      / b / b / b / | / | /1
     :-----------: k / k / k :
     | e | e | e | / | / | /1
     :-----------: k / k /
     | k | e | r | / | /
     :-----------: k /
     | e | e | e | / bottom
     '-----------' done
X11(tv)  .-----------,
        / t / t / t / |
       /---/---/---/ r :
       / t / t / t / | /1
      /---/---/---/ r / r :
      / t / t / t / | / | /1
     :-----------: r / r / r :
     | f | f | f | / | / | /1
     :-----------: f / r /
     | f | f | r | / | /
     :-----------: r /
     | f | f | f | / middle
     '-----------' pos. done
X11(bv)  .-----------,
        / b / k / b / |
       /---/---/---/ k :
       / b / b / b / | /1
      /---/---/---/ k / k :
      / b / b / b / | / | /1
     :-----------: k / k / k :
     | e | e | e | / | / | /1
     :-----------: e / k /
     | e | e | k | / | /
     :-----------: k /
     | e | e | e | / middle
     '-----------' pos. done
```

FIG. X6 showed the CUBE after a R/V/I/I translation on the back 'k' face sent the e,b edge cube to the bottom layer and the back face re-inverted. Then FIG. X6 showed that the other two bottom edge cubes are already in the bottom layer positioned correctly but not oriented correctly so they needed to be bumped back to the middle layer, the b,f edge cube was chosen. FIG. X7 then showed the CUBE after a R/V/I/I translation on the left 'e' face bumped the f,b edge cube to the middle layer. Fig. X7 also showed that by moving the middle layer one Q turn to the right and then inverting the left 'e' face put the b,f edge cube in a READY position to bump back edge cube k,e. FIG. X8 then showed the CUBE after a R/V/I/I translation on the left 'e' face sent the f,b edge cube to the bottom layer and the left face was re-inverted. FIG. X9 then showed the CUBE after a R/V/I/I translation on the right 'r' face bumped the k,b edge cube to the middle layer. FIG. X9 also showed that as for the f,b edge cube, adjusting the middle layer one Q turn to the right and inverting the right 'r' face put the k,b edge cube in a READY position to bump back edge cube k,e. Figure X10 shows the CUBE after a R/V/I/I translation on the right face sends the k,b edge cube to the bottom layer and the right face is re-inverted. Figure X10 shows that the bottom layer is done. Perusal of Figure X10 shows that only middle edge cube f,r is positioned correctly so Figure X11 shows the CUBE after one RM/O operation on the right 'r' face positions correctly all four middle edge cubes. Perusal of Figure X11 shows that diagonally opposite edge cubes f,r and k,e need to be re-oriented by inverting any one of the four faces and

```
X12(tv)  .-----------,
        / t / t / t / |
       /---/---/---/ r :
       / t / t / t / | /1
      /---/---/---/ r / r :
      / t / t / t / | / | /1
     :-----------: r / r / r :
     | f | f | f | / | / | /1
     :-----------: r / r /
     | f | f | f | / | /
     :-----------: r /
     | f | f | f | / CUBE
     '-----------' DONE
X12(bv)  .-----------,
        / b / b / b / |
       /---/---/---/ k :
       / b / b / b / | /1
      /---/---/---/ k / k :
      / b / b / b / | / | /1
     :-----------: k / k / k :
     | e | e | e | / | / | /1
     :-----------: k / k /
     | e | e | e | / | /
     :-----------: k /
     | e | e | e | / CUBE
     '-----------' DONE
```

doing an ALG01 operation that finishes the CUBE after the face is re-inverted. This seventh exercise took 145 steps (188 twists).

EXERCISE-SEVEN SUMMARY

a) Five of the eight top layer cubes are okay after 14 steps moves them from the side of either the middle or bottom layer. One top corner cube was okay after alignment of the top layer and a top edge cube was okay after it was aligned with a top corner cube before a transfer.

b) The remaining top edge cube took seven steps plus an alignment for moving-out and then moving it to the top layer since it was in the top layer incorrectly.

c) Similar to Exercises 4 after the top is done and the middle layer is in alignment two correctly positioned bottom corner cubes are found. Then all four-bottom corner cubes are positioned correctly after a R/O operation interchanges the other two adjacent corners.

d) Like Exercises 4 and 6 none of the four-bottom corner cubes are oriented correctly (are okay). A random R/B/1/1 translation on either of the four faces f, r, k, or e is required. (Table 10EX at the end of this 'exercise' section shows that the necessity for such a random translation on one of the faces occurs four times in these ten exercises.) But unlike Exercises 4 and 6 one corner cube is okay and the other three are already 'in-sync' so a subsequent R/B/1/1 translation (one instead of two) orients all four bottom corner cubes.

e) The four bottom edge cubes are okay after three R/V/1/1 translations move them to the bottom layer after they are 'adjusted' to a READY position. Only three instead of four translations are required since one bottom edge cube was already okay. Similar to Exercise 4, after making two-bottom edge cubes okay the other two were already in the bottom layer (incorrectly) so adjusting the middle layer couldn't put them in a READY position. But unlike Exercise 4 two extra R/V/1/1 translations were required since both edge cubes were in their correct position but needed to be oriented where as for Exercise 4 they were not in their correct position thus sending one of them to the bottom layer automatically bumped the other one back to the middle layer. However, similar to all the previous exercises, before a R/V/1/1 translation a face needed to be inverted (2Qt's) in order to put three of these edge cubes in a READY position. (Table 10EX at the end of this 'exercise' section shows that the necessity for such a bumping-back operation occurs thrice in these ten exercises.)

f) Similar to Exercise 5 only one RM/0 operation, which rotates three of the four middle edge cubes, correctly positions all four middle edge cubes. Two of them are also oriented correctly (are okay).

g) Like Exercise 1, the CUBE is done after an ALG01 operation orients the remaining two middle edge cubes. However, before the operation a face needed to be inverted (2Qt's) in order to make these two edge cubes 'adjacent' (they were on a diagonal) for an ALG01 READY position. (Table 10EX at the end of this 'exercise' section shows that this diagonal re-orientation occurs twice in these ten exercises.)

TABLE X   EXERCISE-SEVEN

OPER-ATION	STEPS	TWISTS	(DO BOTTOM CORNERS)			(bottom edges cont'd)	
			Align	1	--	R/V/1/1(bump)7	3
-----			R/0	9	--	Adjust (1Qt) 1	1
Align	1	--	R/B/1/1	8	--	Invert 'r' 1	--
t,k,e 't' cube	OK	--	R/B/1/1	8	--	R/V/1/1	7	3
Align	1	--	-----			Re-invert 'r' 1	--
t,r,f 't' cube	1	--	Corners Done 26	26		One cube OK --	--
t,r 't' cube	OK	--	=====			-----	
t,f 't' cube	3	2	(DO BOTTOM EDGES)			Bottom Done 44	62
t,r,k 't' cube	3	--	Adjust(2Qt's) 1	1		=====	
t,f,e 't' cube	3	--	Invert 'k' 1	--		(DO MIDDLE EDGES)	
t,e 't' cube	4	1	R/V/1/1	7	3	RM/0	4	2
(t,k) 't' cube	3	2	Re-invert 'k' 1	1		Invert 1	--
t,k 't' cube	4	2	R/V/1/1(bump)7	3		ALG01	45	22
Align	1	--	Adjust (1Qt) 1	1		Re-invert 1	1
-----			Invert 'e' 1	--		-----	
Top Done	24	31	R/V/1/1	7	3	Edges Done 51	75
=====			Re-invert 'e' 1	--		=====	
						CUBE DONE 145	188
						=====	

TABLE X   NOTES

STEPS: A step is one twist of the CUBE that can be one 'Q' turn or two 'Q' turns, e.g. inverting a face is two 'Q' turns.

TWISTS: A twist column shows when a 'step' is two twists instead of one twist, e.g. when bringing down the center slice along with the right side, an extra twist restores the right side; similarly, when rotating the middle slice (layer) with the bottom layer, an extra twist restores the bottom layer. (The vertical slice is not used in solving the CUBE, but is used in pattern generations from a solid CUBE.)

' (prime):   Sometimes at the beginning when doing the top 't' face, I move the bottom layer and middle layer together.

ALIGN: An alignment step aligns the faces of the CUBE (a correction).

(xxx):   A top 't' layer cube that requires two operations since it is either on the bottom 'b' face or incorrectly in the top 't' layer.

ADJUST: A step that places an edge cube, that is to be sent to the bottom 'b' layer, in a READY position.

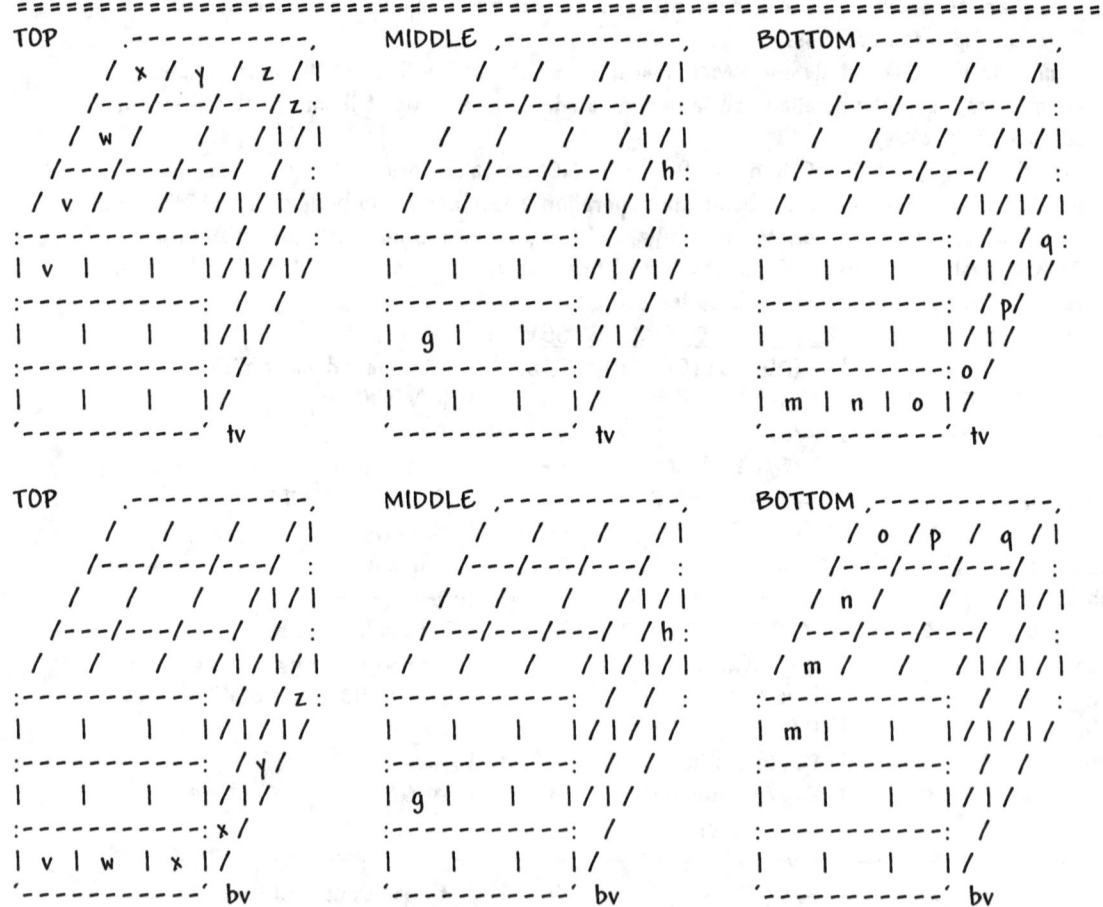

The above three sets of figures show the corresponding corner and edge cubes that share (appear) in both the top (tv) and bottom (bv) views. Of the twenty cubes (not including the six center cubes) twelve do this sharing.

EXAMPLE #8

```
X'1(tv)  .-----------,
        / f / f / r /|
       /---/---/---/ b :
      / b / t / e /|/|
     /---/---/---/ b / e :
    / k / k / e /|/|/|
   :-----------: b / r / e :
   | r | r | k |/|/|/
   :-----------: r / e /
   | k | f | t |/|/
   :-----------: e /
   | r | f | f |/ ELEVEN
   '-----------'   TWISTS

X'1(bv)  .-----------,
        / t / t / t /|
       /---/---/---/ k :
      / t / b / r /|/|
     /---/---/---/ b / f :
    / b / k / b /|/|/|
   :-----------: e / k / k :
   | f | b | f |/|/|/
   :-----------: k / r /
   | e | e | t |/|/
   :-----------: r /
   | t | f | t |/ ELEVEN
   '-----------'   TWISTS

X'3(tv)  .-----------,
        / t / t / t /|
       /---/---/---/ r :
      / t / t / t /|/|
     /---/---/---/ r / b :-e,b
    / t / t / t /|/|/|
   :-----------: r / r / r :
   | f | f | f |/|/|/
   :-----------: f / e /
   | k | f | b |/|/
   :-----------: r /
   | f | f | f |/ corners
   '-----------'   done

X'3(bv)  .-----------,
        / b / k / b /|
       /---/---/---/ k :
      / e / b / b /|/|
     /---/---/---/ k / e :-e,b
    / b / b / b /|/|/|
   :-----------: k / k / k :
   | e | r | e |/|/|/
   :-----------: f / k /
   | r | e | r |/|/
   :-----------: k /
   | e | e | e |/ corners
   '-----------'   done
```

```
X'2(tv)  .-----------,
        / t / t / t /|
       /---/---/---/ r :
      / t / t / t /|/|
     /---/---/---/ r / b :
    / t / t / t /|/|/|
   :-----------: r / r / r :
   | f | f | f |/|/|/
   :-----------: f / k /
   | k | f | b |/|/
   :-----------: b /
   | b | f | f |/ top
   '-----------'   done

X'2(bv)  .-----------,
        / e / b / f /|
       /---/---/---/ b :
      / e / b / k /|/|
     /---/---/---/ e / e :
    / k / b / b /|/|/|
   :-----------: r / k / k :
   | e | r | k |/|/|/
   :-----------: f / k /
   | r | e | r |/|/
   :-----------: k /
   | e | e | e |/ top
   '-----------'   done

X'6(tv)  .-----------,
        / t / t / t /|
       /---/---/---/ r :
      / t / t / t /|/|
     /---/---/---/ r / f :
    / t / t / t /|/|/|
   :-----------: r / r / r :
   | f | f | f |/|/|/
   :-----------: k / r /
   | r | f | e |/|/
   :-----------: r /
   | f | f | f |/ bottom
   '-----------'   done

X'6(bv)  .-----------,
        / b / b / b /|
       /---/---/---/ k :
      / b / b / b /|/|
     /---/---/---/ k / e :
    / b / b / b /|/|/|
   :-----------: k / k / k :
   | e | e | e |/|/|/
   :-----------: r / k /
   | k | e | f |/|/
   :-----------: k /
   | e | e | e |/ bottom
   '-----------'   done
```

Figure X'1 is a scrambled CUBE (after 11 twists with eyes closed). Figure X'2 is the CUBE after moving eight top 't' face cubes via the FOUR STEP APPROACH finishing the top 't' layer. It took a minimum of 16 steps (WOW) plus an alignment since t,e,k needed only to rotate the back 'k' face and t,e,f to rotate the front 'f' face with edge cube t,f as a freebie. Along the way t,r also became a freebie. Finally the middle layer along with the bottom layer were aligned with the top layer. Perusal of Figure X'2 shows that by rotating the bottom layer one Q turn to the left all four bottom corner cubes are positioned correctly (another freebie) plus corner cube b,r,k is okay and Lo-and-Behold the other three corners are already 'in-sync' (WOW, more freebies). Figure X'3 shows the CUBE after two consecutive R/B/1/1 translations on the back 'k' face finishes the orientation of the bottom corner cubes. Perusal of Figure X'3 shows that bottom edge cube b,k is already okay (freebie) and by inverting the back 'k' face edge cube e,b is put in a READY position to bump back edge cube r,b. [FIG. X'4 and FIG. X'5 are not shown but are paraphrased as follows.] FIG. X'4 showed the CUBE after a R/V/1/1 translation on the back 'k' face sent the e,b edge cube to the bottom layer and the back 'k' face is re-inverted. FIG. X'4 also showed that edge cube r,b is already in a READY position to bump back edge cube e,k. Then FIG. X'5 showed the CUBE after a R/V/1/1 translation on the front 'f' face sent the r,b edge cube to the bottom layer. FIG. X'5 also showed that edge cube f,b is already in a READY position to bump back edge cube f,e. Figure X'6 shows the CUBE after a R/V/1/1 translation on the left 'e' face sends the f,b edge cube to the bottom layer. [Amazing,

```
X7(tv)    .-----------,              X'8(tv)    .-----------,
        / t  / t  / t  / l                     / t  / t  / t  / l
      /---/---/---/ r :                       /---/---/---/ r :
      / t  / t  / t  / l / l                  / t  / t  / t  / l / l
    /---/---/---/ r / k :                    /---/---/---/ r / r :
    / t  / t  / t  / l / l / l               / t  / t  / t  / l / l / l
  :-----------: r / r / r :                :-----------: r / r / r :
  l f  l f  l f  l / l / l /               l f  l f  l f  l / l / l /
  :-----------: f / r /                    :-----------: r / r /
  l e  l f  l r  l / l /                   l f  l f  l f  l / l /
  :-----------: r /                        :-----------: r /
  l f  l f  l f  l / middle                l f  l f  l f  l / CUBE
  '-----------' pos, done                  '-----------' DONE
X7(bv)    .-----------,              X'8(bv)    .-----------,
        / b / b / b / l                      / b / b / b / l
      /---/---/---/ k :                      /---/---/---/ k :
      / b / b / b / l / l                    / b / b / b / l / l
    /---/---/---/ k / r :                    /---/---/---/ k / k :
    / b / b / b / l / l / l                  / b / b / b / l / l / l
  :-----------: k / k / k :                :-----------: k / k / k :
  l e  l e  l e  l / l / l /               l e  l e  l e  l / l / l /
  :-----------: e / k /                    :-----------: k / k /
  l f  l e  l k  l / l /                   l e  l e  l e  l / l /
  :-----------: k /                        :-----------: k /
  l e  l e  l e  l / middle                l e  l e  l e  l / CUBE
  '-----------' pos. done                  '-----------' DONE
```

finished the bottom layer edge cubes without adjusting the middle layer.] Perusal of Figure X6 shows that the four middle edge cubes are positioned diagonally opposite so Figure X7 shows the CUBE after doing a C/V/1/2 translation on any face (f, r, k, e). Perusal of Figure X7 shows that all four middle edge cubes are positioned correctly but none of them are oriented, thereby, two ALGO1 operations are required to finish the CUBE; one on any face, the other on the opposite face. This eighth exercise took 98 steps (150 twists).

EXERCISE-EIGHT SUMMARY

a) Four of the eight top 't' layer cubes are okay after 10 steps moved them from the side of either the middle layer or the bottom layer plus one alignment. One top corner was okay after the back 'k' face was rotated. Another top corner and a top edge cube were okay after the front 'f' face was rotated. By happenstance another top edge cube ended up okay. Another edge cube whose face was on the bottom took four steps to move it to the top since edge cubes on the bottom face do not need to be bumped.

b) Like Exercise 4 none of the top 't' layer cubes ended up incorrectly in the top layer nor were corner cubes facing the bottom, therefore, no bumping was needed.

c) Unlike any of the previous exercises the four bottom corners were positioned correctly after the bottom layer was aligned. {Table 10EX at the end of this 'exercise' section shows that this fortuitous happenstance occurs twice in these ten exercises. Here and in Exercise 9.}

d) Like Exercise 2 by happenstance one bottom corner is okay and the other three are already 'in-sync'. Hence, two subsequent R/B/1/1 translations finish the bottom corners (they are all okay).

e) The four bottom edge cubes are okay after three R/V/1/1 translations move them to the bottom layer without any 'adjustment' since by happenstance they were all in a READY position. Only three instead of four translations were required since one bottom edge cube was already okay. However, similar to all the previous exercises, before a R/V/1/1 translation a face needed to be inverted (2Qt's) in order to put one of these edge cubes in a READY position.

f) Similar to Exercise 2 the four middle edge cubes are correctly positioned after a C/V/1/2 translation interchanged them diagonally (instead of along the sides). (Table 10EX at the end of this 'exercise' section shows that this interchange of the middle edge cubes occurs twice in these ten exercises.} However, none of the middle edge cubes are correctly oriented.

g) The CUBE is done after two subsequent ALGO1 operations orient all four middle edge cubes. (Table 10EX at the end of this 'exercise' section shows that on just two occurences this worst case happens when none of the middle edge cubes are oriented correctly.)

TABLE X'   EXERCISE-EIGHT

OPER-ATION	STEPS	TWISTS		(DO BOTTOM CORNERS)			(DO MIDDLE EDGES)	
				Align	1	--	C/V1/2	8	8
				R/B/1/1	8	--	ALG01	45	22
t,e,k 't' cube	1	--		R/B/1/1	8	--	ALG01	45	22
t,e,f 't' cube	1	--		---	---	---	---	---
t,f 't' cube	--	--		Corners Done	17	17	Edges Done	98	150
t,e 't' cube	4	2		===	===	===	===	===
t,r,f 't' cube	4	--		(DO BOTTOM EDGES)			CUBE DONE	155	220
t,r,k 't' cube	3	--		Invert 'k'	1	--
t,r 't' cube	--	--		R/V/1/1	7	3
t,k 't' cube	3	2		Re-invert 'k'	1	--
Align	1	--		R/V/1/1	7	3
				R/V/1/1'	7	3
Top Done	17	21		One cube OK	--	--
				Bottom Done	23	32

TABLE X'   NOTES

STEPS: A step is one twist of the CUBE that can be one 'Q' turn or two 'Q' turns, e.g. inverting a face is two 'Q' turns.

TWISTS: A twist column shows when a 'step' is two twists instead of one twist, e.g, when bringing down the center slice along with the right side, an extra twist restores the right side; similarly, when rotating the middle slice (layer) with the bottom layer, an extra twist restores the bottom layer. (The vertical slice is not used in solving the CUBE, but is used in pattern generations from a solid CUBE.)

' (prime): Sometimes at the beginning when doing the top 't' face, I move the bottom layer and middle layer together.

ALIGN: An alignment step aligns the faces of the CUBE (a correction).

(xxx): A top 't' layer cube that requires two operations since it is either on the bottom 'b' face or incorrectly in the top 't' layer.

ADJUST: A step that places an edge cube, that is to be sent to the bottom 'b' layer, in a READY position.

NOTES:

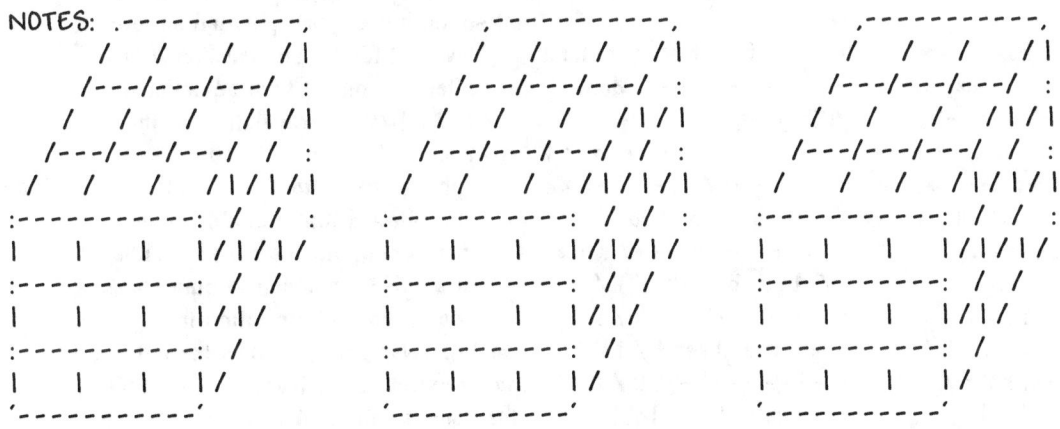

```
Y1(tv)   .-----------,
       / b / b / K /|
      /---/---/---/ t :
     / f / t / f /|/|
    /---/---/---/ t / b :
   / t / r / K /|/|/|
  :-----------: t / r / b :
  | r | t | e |/|/|/
  :-----------: b / K /
  | r | f | e |/|/
  :-----------: f /
  | K | K | t  |/ TWENTY-
  '-----------'TWO TWISTS
```

```
Y1(bv)   .-----------,
       / e / r / r /|
      /---/---/---/ f :
     / t / b / b /|/|
    /---/---/---/ r / K :
   / b / e / e /|/|/|
  :-----------: b / K / r :
  | r | t | K |/|/|/
  :-----------: e / f /
  | f | e | K |/|/
  :-----------: f /
  | f | e | e |/ TWENTY-
  '-----------'TWO TWISTS
```

```
Y5(tv)   .-----------,
       / t / t / t /|
      /---/---/---/ r :
     / t / t / t /|/|
    /---/---/---/ r / e :
   / t / t / t /|/|/|
  :-----------: r / r / r :
  | f | f | f |/|/|/
  :-----------: e / f /
  | r | f | K |/|/
  :-----------: r /
  | f | f | f |/ corners
  '-----------' done
```

```
Y5(bv)   .-----------,
       / b / b / b /|
      /---/---/---/ K :
     / e / b / K /|/|
    /---/---/---/ b / b :
   / b / r / b /|/|/|
  :-----------: K / K / K :
  | e | b | e |/|/|/
  :-----------: K / K /
  | f | e | r |/|/
  :-----------: K /
  | e | e | e |/ corners
  '-----------' done
```

```
Y3(tv)   .-----------,
       / t / t / t /|
      /---/---/---/ r :
     / t / t / t /|/|
    /---/---/---/ r / e :
   / t / t / t /|/|/|
  :-----------: r / r / r :
  | f | f | f |/|/|/
  :-----------: e / f /
  | r | f | K |/|/
  :-----------: b /
  | f | f | r |/ top
  '-----------' done
```

```
Y3(bv)   .-----------,
       / f / e / b /|
      /---/---/---/ K :
     / b / b / r /|/|
    /---/---/---/ b / b :
   / b / K / e /|/|/|
  :-----------: b / K / K :
  | e | b | K |/|/|/
  :-----------: K / K /
  | f | e | r |/|/
  :-----------: K /
  | e | e | e |/ top
  '-----------' done
```

```
Y11(tv)   .-----------,
        / t / t / t /|
       /---/---/---/ r :
      / t / t / t /|/|
     /---/---/---/ r / f :
    / t / t / t /|/|/|
   :-----------: r / r / r :
   | f | f | f |/|/|/
   :-----------: r / r /
   | r | f | K |/|/
   :-----------: r /
   | f | f | f |/ bottom
   '-----------' done
```

```
Y11(bv)   .-----------,
        / b / b / b /|
       /---/---/---/ K :
      / b / b / b /|/|
     /---/---/---/ K / e :
    / b / b / b /|/|/|
   :-----------: K / K / K :
   | e | e | e |/|/|/
   :-----------: K / K /
   | f | e | e |/|/
   :-----------: K /
   | e | e | e |/ bottom
   '-----------' done
```

EXAMPLE #9

Figure Y1 is a scrambled CUBE (after 22 twists with eyes closed). FIG. Y2, not shown here, shows corner cube t,r,f is okay after rotating the top layer one Q turn to the right. Figure Y3 is the CUBE after moving seven top 't' face cubes that were on the sides of either the middle layer or the bottom layer using the Four Step Approach. It took 24 steps to move them correctly to the top 't' layer. It took another seven steps for a bump (t,r edge cube incorrectly in the top layer) in order to finish the top layer. Lo-and-Behold Figure Y3 shows that the four bottom corner cubes are positioned correctly, a freebie, and corner cubes b,r,k and b,e,f are also oriented correctly but they are on a diagonal. Hence, an OOS3 diagram shows that a R/B/1/1 translation on the right 'r' face gets them 'in-sync'. FIG. Y4, not shown here, showed the bottom four corners 'in-sync' with corner cube b,e,k being okay. Figure Y5 shows the CUBE after two R/B/1/1 translations on the left 'e' face finishes the bottom four corner cubes. Perusal of Figure Y5 shows that moving the middle layer one Q turn to the right puts the e,b edge cube in a READY position to bump back edge cube r,b. [Fig. Y6 through FIG. Y10 are not shown but are paraphrased as follows.] FIG. Y6 showed the CUBE after this one Q turn adjustment and a R/V/1/1 translation on the back 'k' face sent the e,b edge cube to the bottom layer. FIG. Y6 also showed that a double adjustment would put the r,b edge cube in a READY position to bump back edge cube f,b by rotating the middle layer one Q turn to the left and by inverting the front 'f' face. FIG. Y7 then showed the CUBE after these two adjustments and a R/V/1/1 translation on the front 'f face sent the r,b edge cube to the bottom layer

```
Y12(tv)  ,-----------,
        / t / t / t /|
       /---/---/---/ r :
      / t / t / t /|/|
     /---/---/---/ r / r :
    / t / t / t /|/|/|
   :-----------: r / r / r :
   | f | f | f |/|/|/|/
   :-----------: f / r /
   | e | f | r |/|/|/
   :-----------: r /
   | f | f | f |/ middle
   '-----------' pos. done
Y12(bv)  ,-----------,
        / b / b / b /|
       /---/---/---/ k :
      / b / b / b /|/|
     /---/---/---/ k / k :
    / b / b / b /|/|/|
   :-----------: k / k / k :
   | e | e | e |/|/|/|/
   :-----------: k / k /
   | f | e | e |/|/|/
   :-----------: k /
   | e | e | e |/ middle
   '-----------' pos. done
```

```
Y13(tv)  ,-----------,
        / t / t / t /|
       /---/---/---/ r :
      / t / t / t /|/|
     /---/---/---/ r / r :
    / t / t / t /|/|/|
   :-----------: r / r / r :
   | f | f | f |/|/|/|/
   :-----------: r / r /
   | f | f | f |/|/|/
   :-----------: r /
   | f | f | f |/ CUBE
   '-----------' DONE
Y13(bv)  ,-----------,
        / b / b / b /|
       /---/---/---/ k :
      / b / b / b /|/|
     /---/---/---/ k / k :
    / b / b / b /|/|/|
   :-----------: k / k / k :
   | e | e | e |/|/|/|/
   :-----------: k / k /
   | e | e | e |/|/|/
   :-----------: k /
   | e | e | e |/ CUBE
   '-----------' DONE
```

and the front 'f' face re-inverted. FIG. Y7 also showed that adjusting the middle layer one Q turn to the right put the f,b edge cube in a READY position to bump back the f,e edge cube. FIG. Y8 then showed the CUBE after this one Q turn adjustment and a R/V/1/1 translation on the left 'e' face sent the f,b cube to the bottom layer. FIG. Y8 also showed that edge cube k,b was already in the bottom layer correctly positioned but not oriented correctly and needed to be bumped back. FIG. Y9 showed the CUBE after a R/V/1/1 translation on the right 'r' face bumped back the k,b edge cube. Fig. Y9 also showed that a double adjustment put the k,b edge cube in a READY position to bump back the e,f edge cube by rotating the middle layer one Q turn to the right and by inverting the right 'r' face. FIG. Y10 showed the CUBE after these two adjustments and a R/V/1/1 translation on the right 'r' face sent the k,b edge cube to the bottom layer and the right 'r' face is re-inverted finishing the bottom layer. Figure Y11 shows the CUBE after the middle layer in FIG. Y10 is aligned two Q turns with the bottom layer done. Perusal of Figure Y11 shows that only middle edge cube e,k is positioned correctly so Figure Y12 shows the CUBE after two RM/0 operations on the left 'e' face rotates the other three middle edge cubes twice which correctly positions all four middle edge cubes. Perusal of Figure Y12 shows that middle edge cubes f,e and f,r are not oriented correctly so that an ALGO1 operation on the front 'f' face finishes the CUBE as shown in Figure Y13. This exercise took 152 steps or 200 twists.

EXERCISE-NINE SUMMARY

a) Seven of the eight top 't' layer cubes are okay after 24 steps moves them from the side of either the bottom or middle layer with one alignment. One top corner cube was okay after an alignment by rotating the top layer one Q turn.

b) The remaining top edge cube took seven steps for moving-out and then moving it to the top layer since it was in the top layer incorrectly.

c) Unlike any of the other exercises, after the top layer was done all four bottom corner cubes were positioned correctly after a one Q turn alignment.

d) Similar to Exercise 1 two of the bottom corner cubes were okay but were also on a diagonal so a R/B/1/1 translation on the front 'f' face dictated by a OOS3 routine got them 'in-sync'. {Table 10EX at the end of this 'exercise' section shows that an 'OOS' (out-of-sync) routine is used four times in these ten exercises when two of the four corners are oriented correctly, namely, OOS1 once EX.4, OOS2 once EX.6, and OOS3 twice EX.1 and EX.9.} After OOS3 gets them 'in-sync' two subsequent R/B/1/1 translations finishes the bottom four-corner cubes.

e) The four bottom edge cubes are okay after three R/V/1/1 translations move them to the bottom layer after they are adjusted to a READY position. Only three instead of four translations are required since one bottom edge cube was already okay. Somewhat similar to Exercises 4 and 7, a bottom layer edge cube was in the bottom layer incorrectly so an extra R/V/1/1 translation was required to bump it back so it could be put in a READY position. However, like all the exercises at least once before the R/V/1/1 translations a face needed to be inverted (2Qt's) in order to put an edge cube in a READY position.

f) Like Exercises 1, 3, 4, and 6 the four middle edge cubes are correctly positioned after two subsequent RM/0 operations rotates twice three of the four middle edge cubes. Two of them are oriented correctly (are okay).

g) The CUBE is done after an ALGO1 operation orients the two remaining adjacent middle edge cubes.

TABLE Y   EXERCISE-NINE

OPER-ATION	STEPS	TWISTS		(DO BOTTOM CORNERS)			(bottom edges cont'd)	
				R/B/1/1	8	--	R/V/1/1(bump) 7	3
Align	1	--		R/B/1/1	8	--	Adjust (1Qt) 1	1
t,r,f 't' cube	OK	--		R/B/1/1	8	--	Invert 'r' 1	--
t,k 't' cube	4'	--		-------------------			R/V/1/1 7	3
t,f 't' cube	4'	--		Corners Done 24		24	Re-invert 'r' 1	--
t,r,k 't' cube	4	--		===================			One cube OK --	--
t,e 't' cube	4	--		(DO BOTTOM EDGES)			-------------------	
Align	1	--		Adjust(1Qt) 1		1	Bottom Done 43	62
t,e,k 't' cube	3	--		R/V/1/1 7		3	===================	
t,e,f 't' cube	3	--		Adjust(1Qt) 1		1	(DO MIDDLE EDGES)	
(t,r) 't' cube	3	2		Invert 'f' 1		--	Align(2Qt's) 1	1
t,r 't' cube	4	2		R/V/1/1 7		3	RM/0 4	2
-------------------				Re-invert 'f' 1		--	RM/0 4	2
Top Done	31	35		Adjust(1Qt) 1		1	ALGO1 45	22
===================				R/V/1/1 7		3	-------------------	
							Edges Done 54	79
							===================	
							CUBE DONE 152	200
							===================	

TABLE Y   NOTES

STEPS: A step is one twist of the CUBE that can be one 'Q' turn or two 'Q' turns, e.g. inverting a face is two 'Q' turns.

TWISTS: A twist column shows when a 'step' is two twists instead of one twist, e.g. when bringing down the center slice along with the right side, an extra twist restores the right side; similarly, when rotating the middle slice (layer) with the bottom layer, an extra twist restores the bottom layer. (The vertical slice is not used in solving the CUBE, but is used in pattern generations from a solid CUBE.)

' (prime): Sometimes at the beginning when doing the top 't' face, I move the bottom layer and middle layer together.

ALIGN: An alignment step aligns the faces of the CUBE (a correction).

(xxx): A top 't' layer cube that requires two operations since it is either on the bottom 'b' face or incorrectly in the top 't' layer.

ADJUST: A step that places an edge cube, that is to be sent to the bottom 'b' layer, in a READY position.

```
Z1(tv)  .-----------,
       / f / e / e /|
      /---/---/---/ k :
     / r / t / r /|/|
    /---/---/---/ f / k :
   / b / f / t /|/|/|/|
  :-----------: r / r / k :
  | f | b | f |/|/|/|/
  :-----------: r / e /
  | t | f | k |/|/|
  :-----------: r / THIRTY-
  | t | f | b |/ THREE
  '-----------' TWISTS

Z1(bv)  .-----------,
       / k / k / t /|
      /---/---/---/ r :
     / t / b / r /|/|
    /---/---/---/ b / t :
   / e / k / e /|/|/|/|
  :-----------: t / k / b :
  | k | b | f |/|/|/|/
  :-----------: b / f /
  | e | e | e |/|/|
  :-----------: e / THIRTY-
  | r | t | b |/ THREE
  '-----------' TWISTS

Z3(tv)  .-----------,
       / t / t / t /|
      /---/---/---/ r :
     / t / t / t /|/|
    /---/---/---/ r / k :
   / t / t / t /|/|/|/|
  :-----------: r / r / b :
  | f | f | f |/|/|/|/
  :-----------: k / e /
  | e | f | e |/|/|
  :-----------: b /
  | e | r | r |/ cor. pos.
  '-----------' done

Z3(bv)  .-----------,
       / f / f / k /|
      /---/---/---/ r :
     / f / b / r /|/|
    /---/---/---/ k / b :
   / f / b / k /|/|/|/|
  :-----------: e / k / k :
  | b | f | b |/|/|/|/
  :-----------: r / k /
  | b | e | b |/|/|
  :-----------: k /
  | e | e | e |/ cor. pos.
  '-----------' done
```

```
Z2(tv)  .-----------,
       / t / t / t /|
      /---/---/---/ r :
     / t / t / t /|/|
    /---/---/---/ r / k :
   / t / t / t /|/|/|/|
  :-----------: r / r / b :
  | f | f | f |/|/|/|/
  :-----------: k / e /
  | e | f | e |/|/|
  :-----------: b /
  | b | f | f |/ top
  '-----------' done

Z2(bv)  .-----------,
       / e / f / k /|-b,r,k
      /---/---/---/ r :
     / r / b / b /|/|
    /---/---/---/ f / b :-k,b
   / f / k / e /|/|/|/|
  :-----------: b / k / k :
  | r | r | k |/|/|/|/
  :-----------: r / k /
  | b | e | b |/|/|
  :-----------: k /
  | e | e | e |/ top
  '-----------' done

Z5(tv)  .-----------,
       / t / t / t /|
      /---/---/---/ r :
     / t / t / t /|/|
    /---/---/---/ r / k :-k,b
   / t / t / t /|/|/|/|
  :-----------: r / r / r :
  | f | f | f |/|/|/|/
  :-----------: k / k /
  | e | f | e |/|/|
  :-----------: r /
  | f | r | f |/ corners
  '-----------' done

Z5(bv)  .-----------,
       / b / r / b /|
      /---/---/---/ k :
     / f / b / b /|/|
    /---/---/---/ f / b :
   / b / f / b /|/|/|/|
  :-----------: k / k / k :
  | e | e | e |/|/|/|/
  :-----------: r / k /
  | b | e | b |/|/|
  :-----------: k /
  | e | e | e |/ corners
  '-----------' done
```

EXAMPLE #10

Figure Z1 is a scrambled CUBE (after 33 twists with eyes closed). Figure Z2 is the CUBE after moving five top 't' face cubes that were on the sides of either the middle layer or the bottom layer using the Four Step Approach. It took 17 steps to move them correctly to the top layer including two alignments. Corner cube t,r,f is already okay, a freebie. It took another ten steps to move edge cube t,f from the top layer (does not require a bump) and to bump corner cube t,r,k from the bottom layer in order to finish the top layer. Perusal of Figure Z2 shows that bottom corner cubes b,r,k and b,e,k are already positioned correctly so that a R/O translation on the back 'k' face interchanges the other two bottom corner cubes which correctly positions all four corner cubes as shown in Figure Z3. Perusal of Figure Z3 shows that not one of the bottom four corner cubes is okay (oriented correctly), so a R/B/1/1 translation is to be done on either of the four faces, the front 'f face is chosen. FIG. Z4, not shown here, showed the CUBE after the translation on the front face and by happenstance the bottom four corner cubes are 'in-sync' with corner cube r,b,k being okay. Figure Z5 shows the CUBE after one R/B/1/1 translation on the back 'k' face finishes the bottom four corner cubes. Perusal of Figure Z5 shows that edge cube k,b is already in a READY position to bump back edge cube f,b. [FIG. Z6 through FIG. Z8 are not shown but are paraphrased as follows.] Fig. Z6 showed the CUBE after a R/V/1/1 translation on the right 'r' face sent the k,b edge cube to the bottom layer. Fig. Z6 also showed that edge cube f,b is already in a READY position to bump back edge cube f,r. FIG. Z7 then showed the CUBE after a R/V/1/1 translation on the left 'e'

```
Z9(tv)   .-----------,
        / t / t / t /|
       /---/---/---/ r :
      / t / t / t /|/|
     /---/---/---/ r / r :
    / t / t / t /|/|/|
   :-----------: r / k / r :
   | f | f | f |/|/|/|
   :-----------: f / r /
   | r | r | e |/|/
   :-----------: r /
   | f | f | f |/ bottom
   '-----------' done

Z9(bv)   .-----------,
        / b / b / b /|
       /---/---/---/ k :
      / b / b / b /|/|
     /---/---/---/ k / k :
    / b / b / b /|/|/|
   :-----------: k / e / k :
   | e | e | e |/|/|/|
   :-----------: k / k /
   | f | f | e |/|/
   :-----------: k /
   | e | e | e |/ bottom
   '-----------' done

Z11(tv)   .-----------,
         / t / t / t /|
        /---/---/---/ r :
       / t / t / t /|/|
      /---/---/---/ r / r :
     / t / t / t /|/|/|
    :-----------: r / r / r :
    | f | f | f |/|/|/|
    :-----------: r / r /
    | f | f | f |/|/
    :-----------: r /
    | f | f | f |/ CUBE
    '-----------' DONE

Z11(bv)   .-----------,
         / b / b / b /|
        /---/---/---/ k :
       / b / b / b /|/|
      /---/---/---/ k / k :
     / b / b / b /|/|/|
    :-----------: k / k / k :
    | e | e | e |/|/|/|
    :-----------: k / k /
    | e | e | e |/|/
    :-----------: k /
    | e | e | e |/ CUBE
    '-----------' DONE
```

```
Z10(tv)   ,-----------,
         / t / t / t /|
        /---/---/---/ r :
       / t / t / t /|/|
      /---/---/---/ r / k :
     / t / t / t /|/|/|
    :-----------: r / r / r :
    | f | f | f |/|/|/|
    :-----------: r / r /
    | f | f | f |/|/
    :-----------: r /
    | f | f | f |/ middle
    '-----------' pos. done

Z10(bv)   ,-----------,
         / b / b / b /|
        /---/---/---/ k :
       / b / b / b /|/|
      /---/---/---/ k / r :
     / b / b / b /|/|/|
    :-----------: k / k / k :
    | e | e | e |/|/|/|
    :-----------: e / k /
    | e | e | k |/|/
    :-----------: k /
    | e | e | e |/ middle
    '-----------' pos. done
```

face sent the f,b edge cube to the bottom layer. FIG. Z7 also show-ed that by adjusting the middle layer 2Qt's edge cube r,b is put in a READY position to bump back edge cube k,r. FIG. Z8 then showed the CUBE after a R/V/1/1 translation on the front 'f' face sent the r,b edge cube to the bottom layer. FIG. Z8 also showed that a double ad-justment put the b,e edge cube in a READY position to bump back edge cube f,e by adjusting the middle layer one Q turn to the right and by inverting the back 'k' face. Figure Z9 shows the CUBE after a R/V/1/1 translation on the back 'k' face sends the b,e cube to the bottom layer and the back face is re-inverted. Perusal of Figure Z9 shows that an alignment of the middle layer one Q turn to the right makes the f,r middle edge cube okay. Therefore, Figure Z10 shows the CUBE after this alignment and after one RM/0 operation on the right 'r' face positions correctly all four middle edge cubes. Perusal of Figure Z10 shows that middle edge cubes f,e and f,r are okay but middle edge cubes k,e and k,r need to be oriented via an ALGO1 operation on the back 'k' face and then the CUBE is done as shown in Figure Z11. This exercise took 126 steps or 170 twists.

EXERCISE-TEN SUMMARY

a) Five of the eight top 't' face cubes are okay after 17 steps move them from the side of either the bottom or middle layer and included two alignments. One top corner cube was already okay, a freebie.

b) The other two top 't' face cubes are okay after 3 steps moves one edge cube whose 't' face was on the bottom face and 7 steps for moving-out and then moving one top corner cube whose 't' face was also on the bottom face.

c) Similar to Exercises 4 and 7 after the top layer is done and the middle layer is in alignment two bottom corner cubes are already okay. Then all four-bottom corner cubes are positioned correctly after a R/0 operation interchanges the other two adjacent bottom corner cubes.

d) Like Exercise 7 none of the four bottom corner cubes are ori-ented correctly (are okay) so a random R/B/1/1 translation was done on one of the four faces. Then one corner cube turned out okay and the other three are already 'in-sync' so a subsequent R/B/1/1 translation (one instead of two) orients all four bottom corner cubes.

e) The four bottom edge cubes are okay after four R/V/1/1 transla-tions move them to the bottom layer after they are 'adjusted' to a READY position. However like many of the previous exercises, before the R/V/1/1 translation a face needed to be inverted (2Qt's) in order to put one of these edge cubes in a READY position.

f) Similar to Exercises 5 and 7 only one RM/0 operation, which rotates three of the four middle edge cubes, correctly positions all four middle edge cubes. Two of them are also oriented correctly (are okay).

g) The CUBE is done after an ALGO1 operation orients the two remaining adjacent middle edge cubes.

TABLE Z   EXERCISE-TEN

OPER-ATION	STEPS	TWISTS	(DO BOTTOM CORNERS)				(DO MIDDLE EDGES)		
			R/0	9	--				
t,r,f 't' cube	OK	--	R/B/1/1	8	--				
t,f 't' cube	3	2	R/B/1/1	8	--				
t,e,k 't' cube	4	--							
t,k 't' cube	--	--	Corners Done	25	25				
Align	1	--							
t,f,e 't' cube	4	--	(DO BOTTOM EDGES)						
t,e 't' cube	3	1	R/V/1/1	7	3				
Align	1	--	R/V/1/1	7	3		(DO MIDDLE EDGES)		
t,r 't' cube	4	2	Adjust(2Qt's)	1	1				
(t,r,k) 't' cube	4	--	R/V/1/1	7	3				
t,r,k 't' cube	3	--	Adjust(1Qt)	1	1		RM/0	4	2
			Invert 'k'	1	--		ALGO1	45	22
Top Done	27	32	R/V/1/1	7	3				
			Re-invert 'k'	1	--		Edges Done	49	73
			Align (1Qt)	1	1				
							CUBE DONE	134	178
			Bottom Done	33	48				

TABLE Z   NOTES

STEPS: A step is one twist of the CUBE that can be one 'Q' turn or two 'Q' turns, e.g. inverting a face is two 'Q' turns.

TWISTS: A twist column shows when a 'step' is two twists instead of one twist, e.g. when bringing down the center slice along with the right side, an extra twist restores the right side; similarly, when rotating the middle slice (layer) with the bottom layer, an extra twist restores the bottom layer. (The vertical slice is not used in solving the CUBE, but is used in pattern generations from a solid CUBE.)

' (prime): Sometimes at the beginning when doing the top 't' face, I move the bottom layer and middle layer together.

ALIGN: An alignment step aligns the faces of the CUBE (a correction).

(xxx): A top 't' layer cube that requires two operations since it is either on the bottom 'b' face or incorrectly in the top 't' layer.

ADJUST: A step that places an edge cube, that is to be sent to the bottom 'b' layer, in a READY position.

TABLE 10EX

path	EX1	EX2	EX3	EX4	EX5	EX6	EX7	EX8	EX9	EX10	eyes-closed
tws----10	10	20	30	40	35	15	25	11	22	33	
i	X	X	X		X	X		X			0 or 1 ?
ii				X			X		X	X	2 or more?
iii								X	X		4 pos. ?
iv		X	X	X	X	X	X			X	R/0
v	X										RR/0
stp	40	31	27	26	30	33	24	17	31	27	cor. pos.
tws	43	33	27	31	34	41	31	21	35	32	done
vi				X		X	X			X	one OK? #
vii	X	X	X	X	X	X	X	X	X	X	in-sync1 #
vii	X	X		X		X			X		in-sync2 #
viii'				X							OOS1 #
viii''						X					OOS2 #
viii'''	X								X		OOS3 #
=========											corns. OK
ix	X	X	X	X	X				X	X	1st bot.
lx	X	X	X	X	X	X	X	X	X	X	2nd edge
lx	X	X	X	X	X	X	X	X	X	X	3rd OK
lx	X	X	X	X	X	X	X	X	X	X	4th *
x											1st B
x											2nd U
x				X			X				3rd M
x							X		X		4th P*
xi	X	X	X								1st inv.
xi	X						X	X	X		2nd inv.
xi		X	X	X	X	X	X				3rd inv.
xi					X		X		X	X	4th inv.
=========											bot. OK
xii	X		X	X	X	X	X		X	X	one pos.?
xiii								X			on diag. @
xiv		X									on side @
xv		X						X			two pos.?
xvi	X		X	X	X	X	X		X	X	RM/0-1
xvi	X		X	X		X			X		RM/0-2
=========											mid. Pos. done
xvii			X		X						four OK?
xviii	X						X				inv. <>
xix		X		X		X		X	X	X	once <>
xix						X		X			twice <>
=========											DONE
steps	164	146	77	162	88	198	145	155	152	134	(1421)
twists	208	192	105	213	110	265	188	220	200	178	(1879)

[average steps = 142 per exercise; average twists = 188 per exercise]

[minimum steps is 77; maximum steps is 198]

= R/B/1/1   * = R/V/1/1   @ = C/V/1/2   <> = ALG01

[average steps top face done = 29 per exercise; average twists top face done = 33 per exercise]

[minimum steps top face is 17; maximum steps top face is 40]

EXAMPLES

How About A "KIBUC" Solution

Here we go from a solid CUBE to a scrambled CUBE by doing the CUBIK solution in reverse, e.g. with regards to Exercise 3, we start with Figure S14'' and end up with Figure S1''. I got this KIBUC idea while red lining (reviewing) my exercise pattern hand written manuscript for Exercise 3. Initially, I indicated that two RM/0 operations on Figure S12'' resulted in a solid CUBE. After a review I decided to insert Figure S13'' showing the CUBE after the first of two subsequent RM/0 operations. In order to generate Figure S13'' and with a few attempts, I had to do two RM/0 operations in reverse to regenerate Figure S12'' before Figure S13'' can be generated.

I decided to choose Exercise 3 in reverse since according to Table 10EX Exercise 3 has the fewest number of steps of the ten exercises.

Now with a little care I was able to get from Figure S14'' to Figure S2'' by doing the CUBIK solution in reverse. Now to get to Figure S1'' from Figure S2'', first I reproduced Figure S1'' using my CUBIK solution so I know it is authentic (verified). Second, I have the eleven entries in Table S'' involving a Four Step Approach that references Figure A to H, however, READY positions may be either one Q turn or two Q turns for the alignments or also for the four steppers (the three steppers are already in a READY position). Therefore, there are five arbitrary READY moves that need to be overcome by trial and error. Now I wish I had made some interim figures or a detailed description during the Four Step Approach noted seven sequences (A to E) for solving the top layer as listed in Table S''. This reverse exercise has been beneficial so far since I discovered a couple of copy errors in the figures. Oh boy, it seems that I have created a puzzle within a puzzle that I'm not able to solve lacking a detailed description of the eleven Table S'' entries for solving the top layer and/or lacking several additional figures.

Consequently, for Exercise 2 I did Figure S15' to Figure S2' in reverse as well as for Exercise 1 doing Figure S24 to Figure S3 in reverse. Since ALGO1 is cyclic in two, doing it forward gets the same result as doing it backward. Figure S1 and Figure S1' are authentic since they were generated using the CUBIK algorithms.

Ten Exercise Errata Section

At this moment I am imagining that a computer 'cubic' program can fill in the missing 'links' in my table entries for solving the top 't' face of the CUBE. Then one is able to go from the 'top done' pattern to the scrambled CUBE pattern. However, the computer most likely has to run forward since it is difficult to conceive of a flowchart or computer program running backwards. This wants me to think about those things in our materialistic world that do run also in reverse, automobiles, trains, bumper cars at amusement parks, model trains with E-units, video projections, unicycles, etc. There are things that go up and down like elevators, but they don't count. Nothing in nature seems to be in reverse even though animals can back up, but that's not really being in reverse.

ERRATUM

Exercise One Patterns: while doing patterns S24 to S3 in reverse and generating pattern S1 using the CUBIK solution I didn't find any mistakes in the 46-pattern manuscript.

Pattern S2 is verified by generating it using the CUBIK solution and then arriving at Pattern S3 using inputs from Table S as follows {I didn't attempt to go from patterns S2 to S1 or from S1 to S2, that's for a computer to do):

a)   Edge cube t,f is READY by moving the bottom layer one Q turn to the left. Bringing down center slice of front face 'f' (1 Qt) and rotating bottom layer 2 Qt's moves in edge cube t,f. Restore top 't' and edge cube t,f is done.

b)   Edge cube t,r is already READY. Bringing down center slice of right 'r' face and rotating bottom layer one Q turn to the right moves in edge cube t,r. Restore top 't' and edge cube t,r is done.

c)   Corner cube t,r,f is READY to be bumped to the side, since it is on the bottom face, by moving bottom layer one Q turn to the left (puts it directly under the t,r,k corner cube). Bringing down the right 'r' slice (front 'f' face towards you) and rotating the bottom layer one Q turn to the right moves/bumps corner cube t,r,f to the side. Restore the right 'r' slice.

d)   Corner cube t,r,f is READY by moving the bottom layer one Q turn to the left. Bringing down the front 'f' slice (right 'r' face towards you) and rotating the bottom layer one Q turn to the left moves in corner cube t,r,f. Restore top 't' and cube t,r,f is done.

e)   Rotating the bottom layer two Q turns puts corner cube t,r,k in a READY position to be bumped to the side like the t,r,f corner cube. Bringing down the back 'k' slice (right 'r' face towards you) and rotating the bottom layer one Q turn to the right moves (bumps) corner cube t,r,k to the side. Restore the back 'k' slice.

f)   Corner cube t,r,k is then READY by moving the bottom layer one Q turn to the left. Bringing down the right 'r' slice (back 'k' face towards you) and rotating the bottom layer one Q turn to the left moves in corner cube t,r,k. Restore the top 't' and corner cube t,r,k is done.

```
PAI(tv)  .-----------,        PAI'(tv)  ,-----------,       Well, it seems like a miracle after all
(altered)/ b / b / b / |             / b / b / b / |       this that the CUBE looks like the S3
       /---/---/---/ f :            /---/---/---/ f :       pattern with the 'top done' which
      / b / t / b / | / |          / b / t / b / | / |      verifies that the pattern S2 is proper
     /---/---/---/ f / f :        /---/---/---/ f / f :     and following the above description
    / b / b / b / | / | / |      / b / b / b / | / | / |    in reverse I can go from pattern S3
   :-----------: f / r / f :    :-----------: f / r / f :   to pattern S2.
   | r | r | r | / | / | /      | r | r | r | / | / | /         Exercise Two Patterns: while doing
   :-----------: f / f /        :-----------: k / f /      patterns S15' to S2' in reverse I didn't
   | r | f | r | / | /      r,f-| r | f | r | / | /        find any mistakes in the 28-pattern man-
   :-----------: f /            :-----------: f /          uscript.  However, generating pattern S1'
   | r | r | r | / CHRIST-      | r | r | r | / not         using the CUBIK solution I discovered
   '-----------' MAN DOTS       '-----------' altered       that it had two t,k edge cubes and by
PAI(bv)  .-----------,        PAI'(bv)  ,-----------,       deduction edge cube r,k was missing.
(altered)/ t / t / t / |             / t / t / t / |       So, which t,k edge cube is the r,k
       /---/---/---/ e :            /---/---/---/ e :       edge cube, a 50-50 proposition.
      / t / b / t / | / |          / t / b / t / | / |      Well I guess wrong.  Hence, when I
     /---/---/---/ e / e :        /---/---/---/ e / e :     got down to the final four middle
    / t / t / t / | / | / |      / t / t / t / | / | / |    edge cubes, I soon realized that two
   :-----------: e / k / e :    :-----------: e / k / e :   of the edge cubes needed to be in-
   | k | k | k | / | / | /      | k | k | k | / | / | /     terchanged.  To illustrate this effect the
   :-----------: e / e /        :-----------: e / e /       patterns shown here show the Christman
   | k | e | k | / | /      r,f-| f | e | k | / | /         Dots patterns with and without an altered
   :-----------: e /            :-----------: e /           CUBE.  Consequently, pattern PAI' shows the
   | k | k | k | / CHRIST-      | k | k | k | / not          effect of altering (interchanging two edge
   '-----------' MAN DOTS       '-----------' altered       cubes) a CUBE without actually doing it.
```

EXAMPLES

Edge cubes r,k and r,f need to be swapped. If any time edge cubes happen to be interchanged, the effect translates down to the last two cubes to be 'solved' needing to be interchanged as depicted in the PA1' pattern above.

Exercise Three Patterns: while doing patterns S15'' to S2'' in reverse I did discover that pattern S4'' had a b,b edge cube in error and it was easy to deduce that it should have been a f,b edge cube. However, generating pattern S1'' using the CUBIK solution I soon discovered that it was a partial disaster since it had four to many 'f's with four less 't's. I did my most intuitive best to try to correct this and authenticated pattern S1'' using my CUBIK solution. The problem is though, pattern S1'' might not be absolutely the original and if so it might be impossible to derive pattern S2'' from pattern S1'' after the 27-steps shown in Table S'' solve (finish) the top 't' layer. Here again I defer to a computer to verify that pattern S2'' can be derived from my resurrected pattern S1'' using the 27-step sequence of Table S''. This problem had no effect on the Table 10EX results, thus the absolute correctness of pattern S1'' is of academic interest only.

Exercise Four Patterns: while doing patterns PS16 to PS2 in reverse I did discover that pattern PS11 had an e,e edge cube and it was easy to deduce that it should have been an f,e edge cube. Also generating pattern PS1 using the CUBIK solution authenticated the pattern.

Exercise Five Patterns: while doing patterns J14 to J3 in reverse I did discover that the bottom view of pattern J3 had the nine letters of the bottom face skewed by ninety degrees (was from top e,b,b / f,b,b / r,k,k; should have been b,b,k / b,b,k / e,f,r). Also generating pattern J1 using the CUBIK solution authenticated the pattern. Pattern J2 is the same as pattern J1 but with the top 't' layer rotated one Q turn. Perhaps a computer can restore the 30-step sequence of Table J for solving the top 't' starting with pattern J2 and ending with pattern J3, then the patterns can all be done in reverse.

Exercise Six Patterns: while doing patterns W14 to W3 in reverse and generating pattern W1 using the CUBIK solution I didn't find any mistakes in the 26-pattern manuscript.

Exercise Seven Patterns: while doing patterns X12 to X2 in reverse and generating pattern X1 using the CUBIK solution I didn't find any mistakes in the 24-pattern manuscript. Perhaps a computer can restore the 24-step sequence of Table X for solving the top 't' starting with pattern X1 and ending with pattern X2, then the patterns can all be done in reverse.

Exercise Eight Patterns: while doing patterns X'8 to X'2 in reverse and generating pattern X'1 using the CUBIK solution I didn't find any mistakes in the 16-pattern manuscript. However, unlike the other nine exercises I was able to restore the 17-step sequence of Table X' for solving the top 't' starting with pattern X'1 and ending with pattern X'2, and then doing the steps in reverse. This happen because the FOUR STEP APPROACH is used only four times while solving the top 't'. These 17 steps are outlined on the next page along with the aid of generating two interim patterns, namely X'1a and X'1b.

```
X'1a(tv)  .-----------,           X'1b(tv)  .-----------,
        / t / r / b / |                   / t / r / b / |
       /---/---/---/ f :                 /---/---/---/ f :
      / t / t / e /|/|                   / t / t / e /|/|
     /---/---/---/ b / t:               /---/---/---/ b / t:
    / t / t / b /|/|/|/|               / t / t / t /|/|/|/|
   :-----------: f / r / t:           :-----------: r / r / b :
   | f | f | r |/|/|/|               | f | f | f |/|/|/|
   :-----------: e / b /             :-----------: r / f /
   | t | f | k |/|/|                 | t | f | f |/|/|
   :-----------: t /                 :-----------: b /
   | k | r | r |/ four               | t | b | r |/ five
   '-----------' 't's done           '-----------' 't's done
X'1a(bv) .-----------,              X'1b(bv) .-----------,
        / k / k / f / |                     / f / b / e / |
       /---/---/---/ r :                   /---/---/---/ k :
      / k / b / f /|/|                     / k / b / k /|/|
     /---/---/---/ r / k:                 /---/---/---/ e / k:
    / e / b / r /|/|/|/|                 / k / k / k /|/|/|/|
   :-----------: k / k / e:             :-----------: b / k / e:
   | b | f | b |/|/|/|                 | r | r | r |/|/|/|
   :-----------: f / b /               :-----------: f / b /
   | r | e | e |/|/|                   | r | e | e |/|/|
   :-----------: k /                   :-----------: k /
   | e | e | e |/ four                 | e | e | e |/ five
   '-----------' 't's done             '-----------' 't's done
```

1) Corner cube t,e,k is done after rotating the back 'k' face two Q turns.

2) Corner cube t,e,f along with edge cube t,f are done after rotating the front 'f' face two Q turns.

3) Edge cube t,e is done via the FOUR STEP APPROACH:
   a) rotating the bottom layer two Q turns puts edge cube t,e in a READY position,
   b) 'bringing down' center 'e' slice one Q turn,
   c) 'moving in' rotating bottom layer two Q turns, and,
   d) 'bringing back' restoring center 'e' slice.
   [Figure X'1a shows the CUBE with the above four top 't' cubes done.]

4) Corner cube t,r,f is done via the FOUR STEP APPROACH:
   a) rotating the bottom layer two Q turns puts corner cube t,r,f in a READY position,
   b) 'bringing down' right slice ('f center facing you),
   c) 'moving in' rotating bottom layer one Q turn to the right,
   d) 'bringing back' restoring right slice.
   [Figure X'1b shows the CUBE with five top 't' cubes done and corner cube t,r,k aligned with edge cube t,r.]

5) Corner cube t,r,k along with edge cube t,r are done together via the FOUR STEP APPROACH:
   a) they are already in a READY position,
   b) 'bringing down' left slice ('k' center facing you),
   c) 'moving in' rotating bottom and middle layers together two Q turns,
   d) 'bringing back' restoring left slice ('f center towards you).

6) Edge cube t,k is done via the FOUR STEP APPROACH:
   a) is already in a READY position,
   b) 'bringing down' center 'f slice,
   c) 'moving in' rotating bottom layer one Q turn to the left,
   d) 'bringing back' restoring center 'f slice.

7) Align bottom layer along with middle layer two Q turns. Top is done as shown in Figure X'2.

Exercise Nine Patterns: while doing patterns Y13 to Y3 in reverse and generating pattern Y1 using the CUBIK solution I didn't find any mistakes in the 24-pattern manuscript. Pattern Y2 is the same as pattern Y1 but with the top layer rotated one Q turn. Perhaps a computer can restore the 31 step sequence of Table Y for solving the top starting with pattern Y2 and ending with pattern Y3, then the patterns can all be done in reverse.

Exercise Ten Patterns: while doing patterns Z11 to Z2 in reverse and generating pattern Z1 using the CUBIK solution I didn't find any mistakes in the 22-pattern manuscript. Perhaps a computer can restore the 27 step sequence of Table Z for solving the top starting with pattern Z1 and ending with pattern Z2, then the patterns can all be done in reverse.

NOTES:

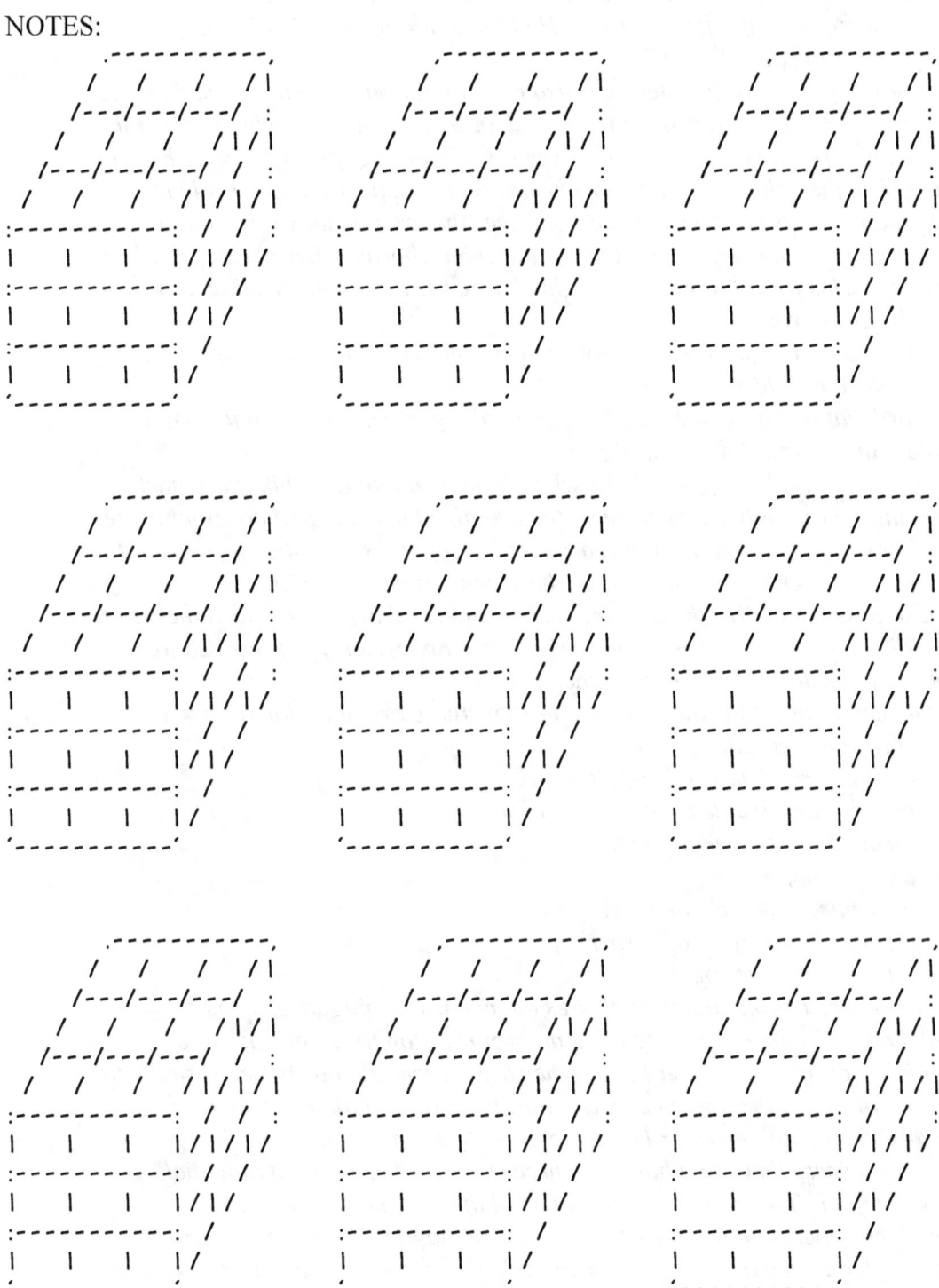

EPILOGUE

Of course my son-in-law is right on target when he alluded to the fact that publishing my 'A CUBIK Year' is just for my own edification since there is just so much 'cubic' information out there on the web sites. He hinted that in lieu of publishing I should open my own web site, NO WAY.

Of course Douglas R. Hofstadter was also right on target when he alluded to the nuances in solving the CUBE in his March 1981 Scientific American article. Our family doctor, to whom my wife sees once a month, never saw the inside of a Rubik's CUBE so I enjoyed showing him the colored front cover, the nine cubic patterns, and the 14 views of the internal mechanism. Over the years I perused over the article many times but never really read it carefully for it contains a lot of minute cubic characteristics which reading can put you asleep. This time though I high lighted my copy those phrases that pertain to solving the CUBE as follows:

*** *"take on the average two weeks of concentrated effort", that's about what it took along with some copious notes;*

*** *"is so hard that it is necessary to find a general algorithm for doing it from any scrambled state", that's what I devised, algorithms;*

*** *"has built up a small science", that's what developed into a 'cubik' flowchart;*

*** *'constraints on the eighth corner cube, the twelfth edge cube, and positioning the last two cubes', boy!, over the last twenty years running into these constraints time and time again with trial and error, it's neat to see these constraints in writing;*

*** *"it is also possible to give three corner cubes one-third twists, provided they are all in the same direction", which is my 'in-sync' routine, and was my first realization when I first started translations on my reference cube 1;*

*** *"what it comes down to is that we have to show there are operators that will perform seven classes of operations:*

(1)   an arbitrary double edge-pair swap,
(2)   an arbitrary double corner-pair swap,
(3)   an arbitrary two-edge flip,
(4)   an arbitrary mason,
(5)   an arbitrary 3-cycle of edges,
(6)   an arbitrary 3-cycle of corner, and,
(7)   an arbitrary baryon,

and these can be reduced to the first four classes but can use conjugate elements".

In my 'cubik' solution after completing an arbitrary top face I next need to position the bottom four corners and if they are incorrectly positioned (usually the case) at least two of the bottom corners must be correct (either adjacent or diagonal). So my R/0 operator being a pseudo '(2)' operation swaps just two of the bottom corners that are adjacent. Three swaps are required if diagonal (however, my RR/0 operator swaps just two of the bottom corners that are diagonal, one operation instead of three, but the diagonal case is very rare). Thereby, I don't need a '(6)' operator to cycle the corners. I next need to orient the bottom four corners and if they are incorrectly oriented (usually the case) they must appear as an arbitrary baryon ('in-sync') or one or two arbitrary mesons (not 'in-sync'). If a baryon, one or two of my reference-1 operators being a '(7)' operator twists three of the four corners until correctly oriented. If masons, a reference-1 operator turns the masons into a baryon so I don't need a '(4)' operator to generate masons. I next need to orient and position correctly the bottom four edge cubes and if any of the four are incorrect (usually the case) I then use my 'bizarre' reference-2 operator to correctly position and orient each bottom edge cube (a minimum of four operations and a maximum of eight operations if none are initially correct). The top and bottom layers are now done. During my R/0, RR/0, and reference-1operations I let the eight middle and bottom edge cubes fall as they may. Similarly, during my reference-2 operations I let the four middle edge cubes fall as they may.

I next need to position the four middle edge cubes and if they are incorrectly positioned (usually the case) either none are correct, one is correct, or all four are correct after being swapped. So my reference-4 operator being a '(1)' operator swaps the pairs of middle edge cubes in order to make either one correct or all four correct. When one is correct my RM/0 operator being a '(5)' operator cycles three middle edge cubes so that one or two successive operations positions them correctly.

Finally, I need to orient the four middle edge cubes and we are done. If they are incorrectly oriented (usually the case for two of them; rarely the case for all four of them) I use me reference-3 operator being a '(3)' operator as part of my ALGO1 to flip a pair of the middle edge cubes (two ALGO1 Operations are needed for the rare case to flip two pairs).

All seven of my operators [reference-1, reference-2, reference-3. reference-4, R/0, (RR/0 optional), RM/0, plus ALGO1] disturb the top layer but restore it as part of the operation. ALGO1 is composed of several reference-2 operations with one reference-3 operation so that the middle and bottom edge cubes are disturbed but are restored at the end of ALGO1.

It seems difficult for me to say that my 'cubik' solution contains any conjugate elements. Two chapters in the Taylor/Rylands book explain a 'method of conjugates'. Do I fool the CUBE, that's the question? My solution is therefore forward going but it is long. Perhaps after all it's the CUBE that is clever, NOT ME!!!

*** *In practice a CUBE solver must develop a fairly large and versatile set of routines that are short, easy to remember, and highly redundant", well my seven routines take seven or eight basic steps and are used over and over again;*

*** *"the typical solver evolves a set of transforms partly by intuition (my seven step sequence), partly by luck (my ALGO1 was luck lucky luck), sometimes with the aid of diagrams --- (my copious notes);*

*** *"most ' cubemeisters' do not think in such detail about how their tools are achieving their goals", that's me;*

*** *"I've forgotten how to solve the CUBE, but my fingers remember", yep, that's it, your sub-conscience takes over, just like playing a solo in front of a band or orchestra;*

*** *"average operator seems to be about 10 to 20 turns long",*

EPILOGUE

Operator	No. of Q turns
Reference-1	8
Reference-2	10
Reference-3	11
Reference-4	16
R/0	9
RR/0 (optional)	10
RM/0	6
ALGO1(five reference-2's plus one reference-3).	67

Of course all of the above cubic nuances are after the fact!

Back in 1962 after I got my M.S as an E.E. I remember taking two post-graduate courses, the first by the physics department in field theory (electrostatics and magnetic fields, etc.) and the second and last by the mathematics department modern (abstract) algebra using the text by Herstein. To my dismay this course was my waterloo, I had hit a brick wall. I remember putting a lot of effort in the course but I didn't get anywhere, I never got it. The instructor was very good and presented the topic as unsophisticated as he could to us eight to ten EE's sitting in the back row "in a fog". At the end of the term he did give us the minimal passing grade (so we could be refunded) for we did turn in our homework but he compassionately told us to not enroll for the second term.

In my own words I often remember the instructor alluding to the fact that sometimes these abstract mathematical concepts would seem realizable to a real world mathematician, then the mathematical scientific researchers may render such a conceptual idea feasible so that (I assume) some kind of model could be constructed.

Now for me to think that it was perhaps very highly probable that such an abstract conceptual idea led to the CUBE is awesome.

By the way, my college has since become a university and these two courses are being taught at a much higher level.

Also, finally, about flowcharting, in the early mid '70's I joined a communications LSI (Large Scale Integration) development group that early on used flowcharting for devising software programs for running (sequencing) essentially command and control type logic systems with some diagnostics. At first the software processing used four-bit machines. However, when the communication systems and processors became more sophisticated, flowcharting soon became too cumbersome and was replaced by a PDL (Program Development Language) for a while, but that soon gave way to higher level languages as these systems and processors just became more and more sophisticated, that's all.

IT HAS BEEN THE 'CUBIK' FLOWCHART THAT I KEPT FOCUSING ON AND IS THE REASON WHY I CONTINUED WORKING ON THIS MANUSCRIPT.

EPILOGUE

```
PS4(tv) .------------,        PS5(tv) .------------,        PS7(tv) .------------,
       / t / t / t /|               / t / t / t /|               / t / t / t /|
      /---/---/---/ r :            /---/---/---/ r :            /---/---/---/ r :
      / t / t / t /|/|             / t / t / t /|/|             / t / t / t /|/|
     /---/---/---/ r / f :        /---/---/---/ r / f :        /---/---/---/ r / e :
     / t / t / t /|/|/|           / t / t / t /|/|/|           / t / t / t /|/|/|
    :------------: r / r / r :-OK :------------: r / r / k :   :------------: r / e / r :
    | f | f | f |/|/|/|           | f | f | f |/|/|/|           | f | f | f |/|/|/|
    :------------: e / f /        :------------: e / b /        :------------: b / k /-k,r
    | k | f | f |/|/|             | k | f | f |/|/|             | b | k | r |/|/|
    :------------: f /            :------------: r /-OK         :------------: r /
    | b | k | b |/ two            | b | f | f |/                | f | f | f |/ b,f & b,r
    '------------' cors. OK       '------------' in-sync        '------------' ready

PS4(bv) .------------,        PS5(bv) .------------,        PS7(bv) .------------,
       / r / r / b /|-OK       OK-/ b / e / r /|                / b / r / b /|
      /---/---/---/ k :            /---/---/---/ b :            /---/---/---/ k :
      / b / b / e /|/|             / r / b / b /|/|             / r / b / e /|/|
     /---/---/---/ b / b :        /---/---/---/ k / b :        /---/---/---/ b / k :
     / e / r / b /|/|/|           / e / r / k /|/|/|           / b / b / b /|/|/|
    :------------: k / k / k :    :------------: e / k / k :    :------------: k / f / k :
    | f | k | e |/|/|/|           | f | k | b |/|/|/|           | e | k | e |/|/|/|
    :------------: r / k /        :------------: r / k /        :------------: f / k /
    | e | e | b |/|/|             | e | e | b |/|/|             | f | r | e |/|/|
    :------------: k /            :------------: k /            :------------: k /
    | e | e | e |/ two            | e | e | e |/                | e | e | e |/ b,f & b,r
    '------------' cors. OK       '------------' in-sync        '------------' ready

PS8(tv) .------------,        PS9(tv) .------------,        PS10(tv) .------------,
       / t / t / t /|               / t / t / t /|                / t / t / t /|
      /---/---/---/ r :            /---/---/---/ r :            /---/---/---/ r :
      / t / t / t /|/|             / t / t / t /|/|             / t / t / t /|/|
     /---/---/---/ r / f :-f,b    /---/---/---/ r / f :        /---/---/---/ r / f :
     / t / t / t /|/|/|           / t / t / t /|/|/|           / t / t / t /|/|/|
    :------------: r / e / r :    :------------: r / r / r :    :------------: r / r / r :
    | f | f | f |/|/|/|           | f | f | f |/|/|/|           | f | f | f |/|/|/|
    :------------: k / r /        :------------: k / r /        :------------: r / r /
    | e | k | e |/|/|             | r | f | e |/|/|        b,e-| e | f | k |/|/|
    :------------: r /            :------------: r /            :------------: r /
    | f | f | f |/                | f | f | f |/                | f | f | f |/
    '------------' b,r done       '------------' b,f done       '------------' bumped b,e

PS8(bv) .------------,        PS9(bv) .------------,        PS10(bv) .------------,
       / b / b / b /|               / b / b / b /|                / b / b / b /|
      /---/---/---/ k :            /---/---/---/ k :            /---/---/---/ k :
   f,r-/ r / b / e /|/|            / b / b / e /|/|             / b / b / r /|/|
     /---/---/---/ b / b :-b,f    /---/---/---/ b / r :        /---/---/---/ f / e :
     / b / b / b /|/|/|           / b / b / b /|/|/|           / b / b / b /|/|/|
    :------------: k / f / k :    :------------: k / k / k :    :------------: k / k / k :
    | e | k | e |/|/|/|           | e | k | e |/|/|/|           | e | k | e |/|/|/|
    :------------: k / k /        :------------: f / k /        :------------: e / k /
    | f | r | r |/|/|             | k | e | e |/|/|        b,e-| b | e | k |/|/|
    :------------: k /            :------------: k /            :------------: k /
    | e | e | e |/                | e | e | e |/                | e | e | e |/
    '------------' b,r done       '------------' b,f done       '------------' bumped b,e
```

Complementary Figures

```
PSII(tv)  .-----------,        PS12(tv)  .-----------,        PS13(tv)  .-----------,
        / b / r / b /|                / t / t / t /|                / t / t / t /|
      /---/---/---/ e :              /---/---/---/ r :              /---/---/---/ r :
      / t / t / t /|/|              / t / t / t /|/|              / t / t / t /|/|
    /---/---/---/ r / r :          /---/---/---/ r / e :          /---/---/---/ r / e :
    / t / t / t /|/|/|            / t / t / t /|/|/|            / t / t / t /|/|/|
  :-----------: r / e / e :      :-----------: r / e / r :      :-----------: r / k / r :
  | f | f | f | f |/|/|/|        | f | f | f | f |/|/|/|        | f | f | f | f |/|/|/|
  :-----------: k / r /          :-----------: b / r /          :-----------: e / r /
  | e | k | e |/|/|              | k | k | k |/|/|              | r | r | k |/|/|
  :-----------: r /              :-----------: r /              :-----------: r /
  | f | f | f |/                 | f | f | f |/                 | f | f | f |/
  '-----------' b,e ready        '-----------' b,e done         '-----------' b,k done

PSII(bv)  .-----------,         PS12(bv)  .-----------,         PS13(bv)  .-----------,
        / b / b / t /|                 / b / b / b /|                 / b / b / b /|
      /---/---/---/ k :               /---/---/---/ k :               /---/---/---/ k :
      / b / b / t /|/|               / b / b / r /|/|               / b / b / b /|/|
    /---/---/---/ k / k :           /---/---/---/ f / f :           /---/---/---/ k / f :
    / b / b / t /|/|/|             / b / b / b /|/|/|             / b / b / b /|/|/|
  :-----------: k / f / k :       :-----------: k / f / k :       :-----------: k / e / k :
  | e | k | r |/|/|/|             | e | e | e |/|/|/|             | e | e | e |/|/|/|
  :-----------: e / f /           :-----------: k / k /           :-----------: r / k /
  | f | r | b |/|/|               | r | r | e |/|/|               | f | f | k |/|/|
  :-----------: k /               :-----------: k /               :-----------: k /
  | e | e | r |/                  | e | e | e |/                  | e | e | e |/
  '-----------' b,e ready         '-----------' b,e done          '-----------' b,k done

J2(tv)  .-----------,            J4(tv  .-----------,            J7(tv)  .-----------,
        / e / b / t /|                 / t / t / t /|                 / t / t / t /|
      /---/---/---/ r :               /---/---/---/ r :               /---/---/---/ r :
      / b / t / t /|/|               / t / t / t /|/|               / t / t / t /|/|
    /---/---/---/ r / f :           /---/---/---/ r / f :           /---/---/---/ r / e :-e,b
    / b / r / e /|/|/|             / t / t / t /|/|/|             / t / t / t /|/|/|
  :-----------: b / r / e :       :-----------: r / r / b :       :-----------: r / f / r :
  | f | f | k |/|/|/|             | f | f | f | f |/|/|/|         | f | f | f |/|/|/|
  :-----------: t / k /           :-----------: k / e /           :-----------: e / r /
  | f | f | e |/|/|               | k | f | r |/|/|               | k | e | f |/|/|
  :-----------: f /               :-----------: r /               :-----------: r /
  | k | k | t |/          b,e,f-| f | r | b |/ two              | f | f | f |/
  '-----------' align top        '-----------'cors. Pos.        '-----------' f,b done

J2(bv)  .-----------,            J4(bv)  .-----------,            J7(bv)  .-----------,
        / e / r / b /|                 / k / k / r /|                 / b / f / b /|
      /---/---/---/ f :               /---/---/---/ f :               /---/---/---/ k :
      / t / b / b /|/|               / b / b / f /|/|               / b / b / r /|/|
    /---/---/---/ r / e :           /---/---/---/ r / e :           /---/---/---/ b / b :-e,b
    / b / e / r /|/|/| b,e,f-/ b / b / e /|/|/|             / b / e / b /|/|/|
  :-----------: f / k / k :       :-----------: b / k / k :       :-----------: k / r / k :
  | r | k | t |/|/|/|             | e | e | k |/|/|/|             | e | k | e |/|/|/|
  :-----------: b / e /           :-----------: f / k /           :-----------: r / k /
  | t | e | f |/|/|               | b | e | b |/|/|               | b | k | k |/|/|
  :-----------: t /               :-----------: k /               :-----------: k /
  | r | k | k |/                  | e | e | e |/ two              | e | e | e |/
  '-----------' align top         '-----------'cors. Pos.         '-----------' f,b done
```

```
J8(tv)   ,------------,        J9(tv)   ,-----------,        J10(tv)  ,-----------,
      / t / t / t / |              / t / t / b / |               / t / t / b / |
    /---/---/---/ r :            /---/---/---/ r :             /---/---/---/ r :
    / t / t / t /|/|             / t / t / f /|/|              / t / t / f /|/|
  /---/---/---/ r / b :-b,k    /---/---/---/ r / k :-b,k     /---/---/---/ r / f :
  / t / t / t /|/|/|           / t / t / b /|/|/|            / t / t / b /|/|/|
 :----------: r / e / r       :----------: r / f / r        :----------: r / f / r
 | f | f | f |/|/|/|           | f | f | k |/|/|/|           | f | f | k |/|/|/|
 :----------: e / r /          :----------: r / r /          :----------: e / r /
 | e | k | k |/|/|             | e | e | k |/|/|       b,r-| r | e | k |/|/|
 :----------: r /              :----------: r /              :----------: r /
 | f | f | f |/|               | f | f | k |/|               | f | f | k |/|
 '----------' e,b done         '----------' b,k ready        '----------' b,k done

J8(bv)   ,------------,        J9(bv)   ,-----------,        J10(bv)  ,-----------,
      / b / f / b / |              / t / t / t / |               / t / t / t / |
    /---/---/---/ k :            /---/---/---/ f :             /---/---/---/ f :
    / b / b / r /|/|             / b / b / r /|/|              / b / b / b /|/|
  /---/---/---/ b / k :-b,k    /---/---/---/ b / b :-b,k     /---/---/---/ k / e :
  / b / b / b /|/|/|           / b / b / b /|/|/|            / b / b / b /|/|/|
 :----------: k / f / k :      :----------: k / r / f :      :----------: k / r / f :
 | e | e | e |/|/|/|           | e | e | e |/|/|/|           | e | e | e |/|/|/|
 :----------: r / k /          :----------: f / k /          :----------: k / k /
 | f | r | k |/|/|             | k | k | e |/|/|       b,r-| b | k | r |/|/|
 :----------: k /              :----------: k /              :----------: k /
 | e | e | e |/|               | e | e | e |/|               | e | e | e |/|
 '----------' e,b done         '----------' b,k ready        '----------' b,k done

J11(tv)  ,------------,        J12(tv   ,-----------,        W2(tv)   ,-----------,
      / t / t / t / |              / t / t / t / |               / e / e / t / |-#
    /---/---/---/ r :            /---/---/---/ r :             /---/---/---/ r :
    / t / t / t /|/|             / t / t / t /|/|              / r / t / k /|/|
  /---/---/---/ r / e :        /---/---/---/ r / f :         /---/---/---/ r / e :
  / b / b / b /|/|/|           / b / b / b /|/|/|            / k / r / f /|/|/|
 :----------: e / f / r        :----------: e / f / r        :----------: e / r / r :
 | f | f | f |/|/|/|           | f | f | f |/|/|/|           | e | b | t |/|/|/|
 :----------: b / r /          :----------: k / r /          :----------: e / k /
 | e | e | r |/|/|             | r | e | e |/|/|             | b | f | k |/|/|
 :----------: e /              :----------: e /              :----------: k /  #-t,r,k
 | f | f | f |/|               | f | f | f |/|               | t | t | r |/|
 '----------' b,r ready        '----------' b,r done         '----------' align t,r,k

J11(bv)  ,-----------,         J12(bv)  ,-----------,         W2(bv)   ,-----------,
      / t / f / b / |              / t / b / b / |               / b / t / f / |
    /---/---/---/ k :            /---/---/---/ k :             /---/---/---/ b :
    / t / b / b /|/|             / t / b / b /|/|              / f / b / e /|/|
  /---/---/---/ k / k :        /---/---/---/ k / e :         /---/---/---/ f / t :
  / t / b / b /|/|/|           / t / b / b /|/|/|            / r / b / e /|/|/|
 :----------: k / r / k :      :----------: k / r / k :      :----------: k / k / k :-#
 | r | e | e |/|/|/|           | r | e | e |/|/|/|           | f | k | t |/|/|/|
 :----------: k / k /          :----------: r / k /          :----------: r / b /
 | f | k | r |/|/|             | k | k | f |/|/|             | f | e | f |/|/|
 :----------: k /              :----------: k /              :----------: b /
 | r | e | e |/|               | r | e | e |/|               | b | t | f |/|
 '----------' b,r ready        '----------' b,r done         '----------' align t,r,k
```

```
W4(tv)   .------------,        W6(tv)   .------------,        W7(tv)   .------------,
        / t / t / t /|                / t / t / t /|                / t / t / t /|
       /---/---/---/ r:               /---/---/---/ r: #-b,r,k      /---/---/---/ r:
       / t / t / t /|/|               / t / t / t /|/|              / t / t / t /|/|
      /---/---/---/ r / b:           /---/---/---/ r / b:          /---/---/---/ r / b:
      / t / t / t /|/|/|             / t / t / t /|/|/|            / t / t / t /|/|/|
     :-----------: r / r / e:        :-----------: r / r / r :-#   :-----------: r / r / K:
     | f | f | f |/|/|/|             | f | f | f |/|/|/|           | f | f | f |/|/|/|
     :-----------: b / K/            :-----------: b / r /         :-----------: b / r /
     | e | f | f |/|/|               | e | f | f |/|/|             | e | f | f |/|/|
     :-----------: r /-b,r,f         :-----------: r /-b,r,f       :-----------: b /
     | (b) | e | f |/  two           | b | e | f |/  two           | b | r | r |/
     '-----------'cors. Pos.         '-----------'cors. OK         '-----------' in-sync

W4(bv)   .------------,        W6(bv)   .------------,        W7(bv)   .------------,
        / b / r / K /|                / b / b / b /|                / f / K / r /|
( )-   /---/---/---/ b:              /---/---/---/ K:              /---/---/---/ b:
b,e,f/ f / b / r /|/|               / f / b / K /|/|              / b / b / f /|/|
      /---/---/---/ b / K:          /---/---/---/ r / K:          /---/---/---/ e / K:
     / (e) / e / r /|/|/|           / e / e / e /|/|/|            / e / e / b /|/|/|
     :-----------: K / K / K:        :-----------: b / K / K:      :-----------: K / K / K:
     | (f) | b | b |/|/|/|           | f | b | K |/|/|/|           | f | b | e |/|/|/|
     :-----------: f / K/            :-----------: f / K/          :-----------: f / K/
     | K | e | r |/|/|               | K | e | r |/|/|             | K | e | r |/|/|
     :-----------: K /               :-----------: K /             :-----------: K / (b,e,k
     | e | e | e |/  two             | e | e | e |/  two           | e | e | e |/     OK)
     '-----------'cors. Pos.         '-----------'cors. OK         '-----------' in-sync

W9(tv)   .------------,        W10(tv)  ,------------,        W11(tv)  ,------------,
        / t / t / t /|                / b / f / b /|                / b / f / b /|
       /---/---/---/ r:              /---/---/---/ e:              /---/---/---/ e:
       / t / t / t /|/|              / t / t / t /|/|              / t / t / t /|/|
      /---/---/---/ r / b :-b,e      /---/---/---/ r / r:          /---/---/---/ r / K:
      / t / t / t /|/|/|             / t / t / t /|/|/|            / t / t / t /|/|/|
     :-----------: r / K / r:        :-----------: r / K / e:      :-----------: r / K / e:
     | f | f | f |/|/|/|             | f | f | f |/|/|/|           | f | f | f |/|/|/|
     :-----------: e / r /           :-----------: e / r /         :-----------: K / r /
     | K | r | K |/|/|               | K | r | K |/|/|             | f | r | r |/|/|
     :-----------: r /               :-----------: r /             :-----------: r /
     | f | f | f |/                  | f | f | f |/                | f | f | f |/
     '-----------' f,b done          '-----------' b,e ready       '-----------' b,e done

W9(bv)   .------------,        W10(bv)  ,------------,        W11(bv)  ,------------,
        / b / b / b /|                / b / b / t /|                / b / b / t /|
       /---/---/---/ K:              /---/---/---/ K:              /---/---/---/ K:
       / b / b / f /|/|              / b / b / t /|/|              / b / b / t /|/|
      /---/---/---/ e / e :-b,e      /---/---/---/ K / f:          /---/---/---/ K / e:
      / b / K / b /|/|/|             / b / K / t /|/|/|            / b / b / t /|/|/|
     :-----------: K / e / K:        :-----------: K / e / K:      :-----------: K / e / K:
     | e | r | e |/|/|/|             | e | r | r |/|/|/|           | e | e | r |/|/|/|
     :-----------: f / K/            :-----------: e / e /         :-----------: b / e /-e,f
     | b | f | r |/|/|               | b | f | b |/|/|             | r | f | K |/|/|
     :-----------: K /               :-----------: K /             :-----------: K / b,e
     | e | e | e |/                  | e | e | r |/                | e | e | r |/ done
     '-----------' f,b done          '-----------' b,e ready       '-----------'(b,K ready)
```

```
X4(tv)   .-----------,            X6(tv)  .-----------,            X7(tv)   .-----------,
        / t / t / t /|                   / t / t / t /|                    / t / t / t /|
       /---/---/---/ r :                /---/---/---/ r :                 /---/---/---/ r :
      / t / t / t /|/|                 / t / t / t /|/|                  / t / t / t /|/|
     /---/---/---/ r / f :            /---/---/---/ r / e :             /---/---/---/ r / b :-b,f
    / t / t / t /|/|/|               / t / t / t /|/|/|                / t / t / t /|/|/|
   :-----------: r / r / b :        :-----------: r / e / r :         :-----------: r / e / r :
   | f | f | f | f |/|/|/           | f | f | f | f |/|/|/            | f | f | f | f |/|/|/
   :-----------: K / r /           :-----------: f / r /             :-----------: r / r /
   | e | f | e |/|/                | e | K | r |/|/                  | f | K | K |/|/
   :-----------: r /-f,r,b         :-----------: r /                 :-----------: r /
   | e | b | f |/ random           | f | b | f |/                   | f | K | f |/
   '-----------' R/B/1/1           '-----------' e,b done            '-----------' bumped b,f

X4(bv)   .-----------,            X6(bv)  ,-----------,            X7(bv)   ,-----------,
        / b / f / K /|                   / b / b / b /|                    / b / b / b /|
       /---/---/---/ r :                /---/---/---/ K :                 /---/---/---/ K :
      / f / b / b /|/|                 / f / b / K /|/|                  / e / b / K /|/|
     /---/---/---/ r / e :            /---/---/---/ b / f :             /---/---/---/ b / f :-b,f
    / f / K / e /|/|/|               / b / b / b /|/|/|                / b / b / b /|/|/|
   :-----------: b / K / K :        :-----------: K / f / K :         :-----------: K / f / K :
   | b | b | K |/|/|/              | e | e | e |/|/|/               | e | e | e |/|/|/
   :-----------: r / K /           :-----------: K / K /             :-----------: e / K /
   | b | e | K |/|/                | K | r | r |/|/                  | r | r | f |/|/
   :-----------: K /               :-----------: K /                 :-----------: K /
   | e | e | e |/ random           | e | e | e |/                   | e | e | e |/
   '-----------' R/B/1/1           '-----------' e,b done            '-----------' bumped b,f

X8(tv)   .-----------,            X9(tv)  ,-----------,            X'4(tv)  ,-----------,
        / t / t / t /|                   / t / t / t /|                    / t / t / t /|
       /---/---/---/ r :                /---/---/---/ r :                 /---/---/---/ r :
      / t / t / t /|/|                 / t / t / t /|/|                  / t / t / t /|/|
     /---/---/---/ r / K :            /---/---/---/ r / f :             /---/---/---/ r / K :
    / t / t / t /|/|/|               / t / t / t /|/|/|                / t / t / t /|/|/|
   :-----------: r / K / r :        :-----------: r / K / r :         :-----------: r / r / r :
   | f | f | f | f |/|/|/           | f | f | f | f |/|/|/            | f | f | f | f |/|/|/
   :-----------: e / r /           :-----------: K / r /             :-----------: b / e /
   | K | r | f |/|/                | (K) | r | r |/|/                | f | f | r |/|/
   :-----------: r /               :-----------: r /                 :-----------: r / e,b
   | f | f | f |/                 | f | f | f |/                    | f | f | f |/ done
   '-----------' f,b done          '-----------' bumped b,K         '-----------'(r,b ready)

X8(bv)   .-----------,            X9(bv)  ,-----------,            X'4(bv)  ,-----------,
        / b / b / b /|                   / b / b / b /|                    / b / K / b /|
       /---/---/---/ K :                /---/---/---/ K :                 /---/---/---/ K :
      / b / b / K /|/|                 / b / b / e /|/|                  / e / b / b /|/|
     /---/---/---/ b / e :            /---/---/---/ K / r :             /---/---/---/ K / r :
    / b / b / b /|/|/|               / b / b / b /|/|/|                / b / b / b /|/|/|
   :-----------: K / e / K :        :-----------: K / e / K :         :-----------: K / K / K :
   | e | e | e |/|/|/              | e | e | e |/|/|/               | e | e | e |/|/|/
   :-----------: f / K /           :-----------: f / K /             :-----------: f / K /
   | r | f | r |/|/                | (b) | f | e |/|/                | r | e | b |/|/
   :-----------: K /               :-----------: K /                 :-----------: K /
   | e | e | e |/                 | e | e | e |/                    | e | e | e |/
   '-----------' f,b done          '-----------' bumped b,K         '-----------' e,b done
```

Complementary Figures

```
X'5(tv)   ,------------,        Y2(tv)   ,------------,        Y4(tv)   ,------------,
        / t / t / t /|                / K / f / K /|                / t / t / t /|
      /---/---/---/ r :            /---/---/---/ e :            /---/---/---/ r :
      / t / t / t /|/|             / b / t / r /|/|             / t / t / t /|/|
    /---/---/---/ r / r :        /---/---/---/ t / b :        /---/---/---/ r / e :
    / t / t / t /|/|/|           / b / f / t /|/|/|           / t / t / t /|/|/|
   :------------: r / r / r :   :------------: r / r / b :   :------------: r / r / K :
   | f | f | f |/|/|/|          | e | e | f |/|/|/|          | f | f | f |/|/|/|
   :------------: r / r /-r,b   :------------: b / K /       :------------: e / b /
   | b | f | K |/|/|            | r | f | e |/|/|            | r | f | K |/|/|
   :------------: r /          :------------: f /          :------------: b /
   | f | f | f |/|             | K | K | t |/ align,        | b | f | r |/|
   '------------' r,b done     '------------' t,r,f OK      '------------' in-sync

X'5(bv)   ,------------,        Y2(bv)   ,------------,        Y4(bv)   ,------------,
        / b / b / b /|                / e / r / r /|                / f / r / r /|
      /---/---/---/ K :            /---/---/---/ f :            /---/---/---/ b :
      / e / b / b /|/|             / t / b / b /|/|             / b / b / K /|/|
    /---/---/---/ K / f :        /---/---/---/ r / K :        /---/---/---/ b / b :
    / b / b / b /|/|/|           / b / e / e /|/|/|           / e / e / b /|/|/|
   :------------: K / K / K :   :------------: b / K / t :   :------------: K / K / K :
   | e | e | e |/|/|/|          | r | t | K |/|/|/|          | f | f | e |/|/|/|
   :------------: e / K /       :------------: e / t /       :------------: K / K /
   | f | e | K |/|/|            | f | e | K |/|/|            | f | e | r |/|/|
   :------------: K /          :------------: t /          :------------: K / (b,e,K
   | e | e | e |/              | f | f | r |/ align,        | e | e | e |/      OK)
   '------------' r,b done     '------------' t,r,f OK      '------------' in-sync

Y6(tv)   ,------------,         Y7(tv)   ,------------,        Y8(tv)   ,------------,
        / t / t / t /|                / t / t / t /|                / t / t / t /|
      /---/---/---/ r :            /---/---/---/ r :            /---/---/---/ r :
      / t / t / t /|/|             / t / t / t /|/|             / t / t / t /|/|
    /---/---/---/ r / f :        /---/---/---/ r / r :        /---/---/---/ r / f :
    / t / t / t /|/|/|           / t / t / t /|/|/|           / t / t / t /|/|/|
   :------------: r / f / r :   :------------: r / r / r :   :------------: r / f / r :
   | f | f | f |/|/|/|          | f | f | f |/|/|/|          | f | f | f |/|/|/|
   :------------: r / f /       :------------: K / r /       :------------: f / r /
   | e | e | b |/|/|            | K | f | e |/|/|            | K | e | r |/|/|
   :------------: r /          :------------: r /          :------------: r /
   | f | f | f |/              | f | f | f |/              | f | f | f |/
   '------------' e,b done     '------------' r,b done     '------------' f,b done

Y6(bv)   ,------------,         Y7(bv)   ,------------,        Y8(bv)   ,------------,
        / b / b / b /|                / b / b / b /|                / b / K / b /|
      /---/---/---/ K :            /---/---/---/ K :            /---/---/---/ K :
      / e / b / K /|/|             / e / b / K /|/|             / b / b / K /|/|
    /---/---/---/ b / r :        /---/---/---/ b / f :        /---/---/---/ b / e :
    / b / b / b /|/|/|           / b / b / b /|/|/|           / b / b / b /|/|/|
   :------------: K / r / K :   :------------: K / K / K :   :------------: K / r / K :
   | e | e | e |/|/|/|          | e | e | e |/|/|/|          | e | e | e |/|/|/|
   :------------: K / K /       :------------: f / K /       :------------: e / K /
   | K | K | r |/|/|            | r | e | b |/|/|            | r | K | K |/|/|
   :------------: K /          :------------: K /          :------------: K /
   | e | e | e |/              | e | e | e |/              | e | e | e |/
   '------------' e,b done     '------------' r,b done     '------------' f,b done
```

```
Y9(tv)    .-----------,              Y10(tv)   .-----------,            Z4(tv)    .-----------,
         / t / t / t /|                      / t / t / t /|                     / t / t / t /|
        /---/---/---/ r :                   /---/---/---/ r :                  /---/---/---/ r :
       / t / t / t /|/|                    / t / t / t /|/|                   / t / t / t /|/|
      /---/---/---/ r / r :               /---/---/---/ r / f :              /---/---/---/ r / k :
     / t / t / t /|/|/|               / t / t / t /|/|/|              / t / t / t /|/|/|
    :-----------: r / f / r :             :-----------: r / e / r :          :-----------: r / r / r :-#
    | f | f | f |/|/|/|                    | f | f | f |/|/|/|                 | f | f | f |/|/|/|
    :-----------: k / r /                  :-----------: e / r /              :-----------: k / k /
    | k | e | r |/|/|                      | e | k | k |/|/|                  | e | f | e |/|/| #-r,b,k
    :-----------: r /                      :-----------: r /                  :-----------: f /
    | f | f | f |/                         | f | f | f |/                     | e | e | b |/
    '-----------' bumped b,k               '-----------' b,k done             '-----------' in-sync
Y9(bv)    .-----------,              Y10(bv)   .-----------,            Z4(bv)    .-----------,
         / b / b / b /|                      / b / b / b /|                     / r / r / b /|-#
        /---/---/---/ k :                   /---/---/---/ k :                  /---/---/---/ k :
       / b / b / e /|/|                    / b / b / b /|/|                   / f / b / f /|/|
      /---/---/---/ f / k :               /---/---/---/ k / r :              /---/---/---/ r / b :
     / b / b / b /|/|/|               / b / b / b /|/|/|              / f / b / e /|/|/|
    :-----------: k / r / k :             :-----------: k / f / k :          :-----------: b / k / k :
    | e | e | e |/|/|/|                    | e | e | e |/|/|/|                 | b | f | k |/|/|/|
    :-----------: r / k /                  :-----------: k / k /              :-----------: r / k /
    | b | k | f |/|/|                      | f | r | r |/|/|                  | b | e | b |/|/|
    :-----------: k /                      :-----------: k /                  :-----------: k / (r,b,k
    | e | e | e |/                         | e | e | e |/                     | e | e | e |/    OK)
    '-----------' bumped b,k               '-----------' b,k done             '-----------' in-sync
Z6(tv)    .-----------,              Z7(tv)    .-----------,            Z8(tv)    .-----------,
         / t / t / t /|                      / t / t / t /|                     / t / t / t /|
        /---/---/---/ r :                   /---/---/---/ r :                  /---/---/---/ r :
       / t / t / t /|/|                    / t / t / t /|/|                   / t / t / t /|/|
      /---/---/---/ r / r :               /---/---/---/ r / r :              /---/---/---/ r / r :
     / t / t / t /|/|/|               / t / t / t /|/|/|              / t / t / t /|/|/|
    :-----------: r / r / r :             :-----------: r / r / r :          :-----------: r / e / r :
    | f | f | f |/|/|/|                    | f | f | f |/|/|/|                 | f | f | f |/|/|/|
    :-----------: e / k /                  :-----------: k / k /              :-----------: e / r /
    | b | f | b |/|/|                      | e | f | e |/|/|                  | k | k | b |/|/|
    :-----------: r /                      :-----------: r /                  :-----------: r /
    | f | r | f |/                         | f | f | f |/                     | f | f | f |/
    '-----------' k,b done                 '-----------' f,b done             '-----------' r,b done
Z6(bv)    .-----------,              Z7(bv)    .-----------,            Z8(bv)    .-----------,
         / b / r / b /|                      / b / r / b /|                     / b / b / b /|
        /---/---/---/ k :                   /---/---/---/ k :                  /---/---/---/ k :
       / f / b / b /|/|                    / b / b / b /|/|                   / b / b / b /|/|
      /---/---/---/ k / b :               /---/---/---/ k / f :              /---/---/---/ k / f :
     / b / f / b /|/|/|               / b / f / b /|/|/|              / b / f / b /|/|/|
    :-----------: k / k / k :             :-----------: k / k / k :          :-----------: k / f / k :
    | e | e | e |/|/|/|                    | e | e | e |/|/|/|                 | e | e | e |/|/|/|
    :-----------: e / k /                  :-----------: r / k /              :-----------: k / k /
    | f | e | k |/|/|                      | b | e | b |/|/|                  | e | r | r |/|/|
    :-----------: k /                      :-----------: k /                  :-----------: k /
    | e | e | e |/                         | e | e | e |/                     | e | e | e |/
    '-----------' k,b done                 '-----------' f,b done             '-----------' r,b done
```

```
la(tv)    .------------,          Cube/Op./Tr.    Stp/Tws      lc(tv)    .-----------,
         / k / k/ k /|            t,f,r (OK)       --/--              / t / t / t /|
        /---/---/---/ e :         t,e              3 / 2            /---/---/---/ r :
       / k / t / r /|/|           t,f              3 / 1           / t / t / t /|/|
      /---/---/---/ f/t:          Align (1Qt)      1'/--          /---/---/---/ r/f:
     / b / b/ t /|/|/|            t,k              4 / 2         / t / t / t /|/|/|
    :-----------: r/r/f:          t,r              4 / 1        :-----------: r/r/r:
    | f | e | f |/|/|/|           Align (1Qt)      1'/--        | f | f | f |/|/|/|
    :-----------: f/ t/           (t,f,e)          3 /--        :-----------: e/b/
    | f | f| b |/|/               t,f,e            3 /--        | k | f | k |/|/
    :-----------: k/              (t,k,r)          3 /--        :-----------: r/
    | k | r| r |/TWELVE           t,k,r            3 /--        | f | b| f |/ corners
    '-----------'  TWISTS         t,k,e            3 /--        '-----------'  done

la(bv)    .------------,          --------------------          lc(bv)    .-----------,
         / b / k / e /|           top done (#)     31 / 37             / b / f / b /|
        /---/---/---/ b|          ====================             /---/---/---/ k :
       / b / b / e /|/|           R/O ('f face)    9 /--           / k / b / e /|/|
      /---/---/---/ t /r:         (*)R/B/1/1 ('r' face)  8 /--     /---/---/---/ f/r:
     / e / b / t /|/|/|           R/B/1/1 ('e' face)  8 /--       / b / e / b /|/|/|
    :-----------: f/k/t:          --------------------          :-----------: k/k/k:
    | b | k | e |/|/|/|           corners done     25 / 25      | e | b | e |/|/|/|
    :-----------: t /r/           ====================          :-----------: r/k/
    | e | e | f |/|/              r,b (rdy, 2Qt's)  1 / 1       | r | e | b |/|/
    :-----------: t/              R/V/1/1 ('f face)  7 / 3      :-----------: k/
    | r | e | r |/TWELVE          f,b (rdy, Inv 'e')  1 /--     | e | e | e |/ corners
    '-----------'  TWISTS         R/V/1/1 ('e' face)  7 / 3     '-----------'  done

lb(tv)    .------------,          Re-inv 'e'        1 /--       ld(tv)    .-----------,
         / t / t / t /|           Adjust (-1Qt)     1 / 1             / t / t / t /|
        /---/---/---/ r:          k,b (rdy, Inv 'r')  1 /--        /---/---/---/ r :
       / t / t / t /|/|           R/V/1/1 ('r' face)  7 / 3       / t / t / t /|/|
      /---/---/---/ r/f:          Re-inv 'r'        1 /--         /---/---/---/ r/r:
     / t / t / t /|/|/|           [R/V/1/1 ('k' face)]  7 / 3    / t / t / t /|/|/|
    :-----------: r/r/b:          Adjust (+1Qt)     1 / 1        :-----------: r/r/r:
    | f | f| f |/|/|/|            e,b (rdy, Inv 'k)  1 /--       | f | f | f |/|/|/|
    :-----------: e/e/            R/V/1/1 ('k' face)  7 / 3      :-----------: k/r/
    | k | f| k |/|/               Re-inv 'k'        1 /--        | f | f| e |/|/
    :-----------: r/              Align (2Qt's)     1 / 1        :-----------: r/
    | b | b| f |/ top             --------------------          | f | f| f |/ bottom
    '-----------'  done           bottom done (z)   45 / 64      '-----------'  done

lb(bv)    .-----------,           ====================          ld(bv)    .-----------,
         / b / b / e /|           RM/O ('f face)    4 / 2             / b / b / b /|
        /---/---/---/ k:          ALGO1 ('k' face)  45 / 22        /---/---/---/ k :
       / f / b / b /|/|           --------------------           / b / b / b /|/|
      /---/---/---/ k/r:          edges done        49 / 73       /---/---/---/ k/f:
     / e / e / b /|/|/|           ====================           / b / b / b /|/|/|
    :-----------: r/k/k:          CUBE DONE        150 / 199     :-----------: k/k/k:
    | f | f| k |/|/|/|            ====================           | e | e | e |/|/|/|
    :-----------: r/k/            (#) f,r,b and f,e,b in position :-----------: r/k/
    | r | e | b |/|/              (*) see page (17) 'in-sync'     | e | e | k |/|/
    :-----------: k/              (z) f,e is OK                   :-----------: k/
    | e | e | e |/ top                                           | e | e | e |/ bottom
    '-----------'  done                  Example 11              '-----------'  done
```

```
11a(tv)    .------------,
          / r / f / K /|
         /---/---/---/ b :
        / f / t / t /|/|
       /---/---/---/ e / b :
      / t / K / t /|/|/|
     :----------- r / r / b :
     | f | t | f |/|/|/|
     :-----------: f/ r/
     | e | f | e |/|/
     :-----------: K /
     | r | r | r |/THIRTEEN
     '-----------'   TWISTS

11a(bv)    .------------,
          / b / t / f /|
         /---/---/---/ e |
        / b / b / K /|/|
       /---/---/---/ r / e :
      / b / f / e /|/|/|
     :-----------: K /K / e :
     | f | t  | t |/|/|/|
     :-----------: b / r /
     | K | e | K |/|/
     :-----------: K /
     | e | b | t |/THIRTEEN
     '-----------'   TWISTS

11b(tv)    .------------,
          / t / t / t /|
         /---/---/---/ r :
        / t / t / t /|/|
       /---/---/---/ r / r :
      / t / t / t /|/|/|
     :-----------: r / r / K :
     | f | f | f |/|/|/|
     :-----------: f / b /
     | e | f | e |/|/
     :-----------: b /
     | r | b | f |/ top
     '-----------'   done

11b(bv)    .------------,
          / e / K / b /|
         /---/---/---/ e :
        / e / b / r /|/|
       /---/---/---/ b / f :
      / K / b / b /|/|/|
     :-----------: f/K / K :
     | b | f | r |/|/|/|
     :-----------: r / K /
     | K | e | K |/|/
     :-----------: K /
     | e | e | e |/ top
     '-----------'   done
```

Cube/Op./Tr.	Stp/Tws
t,f,e (OK)	--/--
t,f,r (OK)	--/--
t,r	3 / 2
t,f	3 / 2
t,e	3 / 2
t,K	4 / 2
t,K,e	3 /--
t,K,r	3 /--
-----------------------	
top done	19 / 27
=======================	
Align (2Qt's) (#)	1 /--
(*) R/0 ('r' face)	9 /--
R/B/1/1 ('r' face)	8 /--
R/B/1/1 ('r' face)	8 /--
-----------------------	
corners done	26 / 26
=======================	
[all edge cubes on bottom]	
[R/V/1/1 ('f face)]	7 / 3
Adjust (-1Qt)	1 / 1
K,b (rdy)	--/--
R/V/1/1 ('r' face)	7 / 3
Adjust (-1Qt)	1 / 1
e,b (rdy)	--/--
R/V/1/1 ('K' face)	7 / 3
Adjust (-1Qt)	1 / 1
f,b (rdy)	--/--
R/V/1/1 ('e' face)	7 / 3
Adjust (2Qt's)	1 / 1
r,b (rdy, Inv 'f')	1 /--
R/V/1/1 ('f' face)	7 / 3
Re-inv 'f'	1 /--
Align (+1Qt)	1 / 1
-----------------------	
bottom done	42 / 61
=======================	
RM/0 ('f' face)	4 / 2
RM/0 ('f' face)	4 / 2
-----------------------	
edges done	8 / 12
=======================	
CUBE DONE	95 / 126
=======================	
(#) r,f,b and r,K,b in position	
(*) r,f,b is OK	

Example 12

```
11c(tv)    .------------,
          / t / t / t /|
         /---/---/---/ r :
        / t / t / t /|/|
       /---/---/---/ r / r :
      / t / t / t /|/|/|
     :-----------: r / r / r :
     | f | f | f |/|/|/|
     :-----------: f / K /
     | e | f | e |/|/
     :-----------: r /
     | f | b | f |/ corners
     '-----------'   done

11c(bv)    .------------,
          / b / b / b /|
         /---/---/---/ K :
        / r / b / b /|/|
       /---/---/---/ e / f :
      / b / b / b /|/|/|
     :-----------: K /K / K :
     | e | f | e |/|/|/|
     :-----------: r / K /
     | K | e | K |/|/
     :-----------: K /
     | e | e | e |/ corners
     '-----------'   done

11d(tv)    .------------,
          / t / t / t /|
         /---/---/---/ r :
        / t / t / t /|/|
       /---/---/---/ r / e :
      / t / t / t /|/|/|
     :-----------: r / r / r :
     | f | f | f |/|/|/|
     :-----------: r / r /
     | f | f | K |/|/
     :-----------: r /
     | f | f | f |/ bottom
     '-----------'   done

11d(bv)    .------------,
          / b / b / b /|
         /---/---/---/ K :
        / b / b / b /|/|
       /---/---/---/ K / K :
      / b / b / b /|/|/|
     :-----------: K /K / K :
     | e | e | e |/|/|/|
     :-----------: f / K /
     | e | e | r |/|/
     :-----------: K /
     | e | e | e |/ bottom
     '-----------'   done
```

ADDENDUM

```
IIIa(tv)    .-----------,
         / K / K / e / |
        /---/---/---/ f :
       / t / t / e / | /|
      /---/---/---/ f / r :
     / r / e / t / | /|/ |
    :-----------: e / r / r :
    | t | b | K | /|/|/|/
    :-----------: t / r /
    | t | f | e | /|/ /
    :-----------: K /
    | f | t | r | / FOURTEEN
    '-----------'   TWISTS

IIIa(bv)  .-----------,
         / b / f / f / |
        /---/---/---/ b |
       / r / b / e / | /|
      /---/---/---/ K / K :
     / b / f / t / | /|/ |
    :-----------: r / K / t :
    | e | b | f | /|/|/|/
    :-----------: r / b /
    | K | e | b | /|/ /
    :-----------: b /
    | K | f | e | / FOURTEEN
    '-----------'   TWISTS

IIIb(tv)  .-----------,
         / t / t / t / |
        /---/---/---/ r :
       / t / t / t / | /|
      /---/---/---/ r / K :
     / t / t / t / | /|/ |
    :-----------: r / r / r :
    | f | f | f | /|/|/|/
    :-----------: f / b /
    | b | f | e | /|/ /
    :-----------: f /
    | K | f | e | / top
    '-----------'  done

IIIb(bv)  .-----------,
         / b / K / b / |
        /---/---/---/ K :
       / r / b / e / | /|
      /---/---/---/ b / e :
     / e / f / f / | /|/ |
    :-----------: r / K / K :
    | b | b | b | /|/|/|/
    :-----------: K / K /
    | r | e | r | /|/ /
    :-----------: K /
    | e | e | e | / top
    '-----------'  done
```

Cube/Op./Tr.	Stp/Tws
Align top (2Qt's)	1/--
t,k,e (OK)	--/--
t,e	3/1
t,K	3/2
t,e'f	3/--
t,f,r	4/--
t,r	3/2
t,f	4/2
(t,r,K)	3/--
t,r,K	4/--
Align (1Qt)	1/--
-----------------	
top done	29/36
=================	
Align (-1Qt) (#)	1/--
(*) R/O ('e' face)	9/--
R/B/1/1 ('K' face)	8/--
-----------------	
corners done	18/18
=================	
Adjust (+1Qt)	1/1
r,b (rdy)	--/--
R/V/1/1 ('f face)	7/3
e,b (rdy)	--/--
R/V/1/1 ('K' face)	7/3
k,b (OK)	--/--
[R/V/1/1 ('e' face)]	7/3
Adjust (+1Qt)	1/1
f,b (rdy, Inv 'e')	1/--
R/V/1/1 ('e' face)	7/3
Re-inv 'e'	1/--
Align (2Qt's)	1/1
-----------------	
bottom done	33/48
=================	
RM/O ('r' face)	4/2
RM/O ('r' face)	4/2
ALG01 ('r' face)	45/22
-----------------	
edges done	53/79
=================	
CUBE DONE	133/181
=================	
(#) e,k,b and e,f,b in position	
(*) k,r,b is OK	

Example 13

```
IIIc(tv)    ,-----------,
         / t / t / t / |
        /---/---/---/ r :
       / t / t / t / | /|
      /---/---/---/ r / K :
     / t / t / t / | /|/ |
    :-----------: r / r / r :
    | f | f | f | /|/|/|/
    :-----------: f / e /
    | b | f | e | /|/ /
    :-----------: r /
    | f | b | f | / corners
    '-----------'  done

IIIc(bv)  ,-----------,
         / b / b / b / |
        /---/---/---/ K :
       / f / b / b / | /|
      /---/---/---/ K / e :
     / b / r / b / | /|/ |
    :-----------: K / K / K :
    | e | f | e | /|/|/|/
    :-----------: K / K /
    | r | e | r | /|/ /
    :-----------: K /
    | e | e | e | / corners
    '-----------'  done

IIId(tv)  ,-----------,
         / t / t / t / |
        /---/---/---/ r :
       / t / t / t / | /|
      /---/---/---/ r / e :
     / t / t / t / | /|/ |
    :-----------: r / r / r :
    | f | f | f | /|/|/|/
    :-----------: f / r /
    | r | f | r | /|/ /
    :-----------: r /
    | f | f | f | / bottom
    '-----------'  done

IIId(bv)  ,-----------,
         / b / b / b / |
        /---/---/---/ K :
       / b / b / b / | /|
      /---/---/---/ K / K :
     / b / b / b / | /|/ |
    :-----------: K / K / K :
    | e | e | e | /|/|/|/
    :-----------: f / K /
    | K | e | e | /|/ /
    :-----------: K /
    | e | e | e | / bottom
    '-----------'  done
```

ADDENDUM

```
IVa(tv)   .- - - - - - - - - - -,          Cube/Op./Tr.        Stp/Tws        IVc(tv)   ,- - - - - - - - - - -,
        / f / e / b / l           Align top (-1Qt)     1 / - -                  / t / t / t / l
      /- - -/- - -/- - -/ f :        t,f,e (OK)         - -/ - -               /- - -/- - -/- - -/ r :
      / f / t / K / l / l             t,f               3 / 2                    / t / t / t / l / l
     /- - -/- - -/- - -/ e / b :      t,k,e             2 / - -                 /- - -/- - -/- - -/ r / e :
    / r / K / t / l / l / l           t,K (OK)          - -/ - -                / t / t / t / l / l / l
    :- - - - - - - - - -: f / r / t :  t,k,r            3 / - -                :- - - - - - - - - -: r / r / r :
    l K l b l e l / l / l             t,f,r             3 / - -                l f l f l f l / l / l
    :- - - - - - - - - -: r / t /      t,r              3 / 1                  :- - - - - - - - - -: b / f /
    l t l f l b l / l /               t,e               4 / 1                  l e l f l f l / l /
    :- - - - - - - - - -: K /                                                  :- - - - - - - - - -: r /
    l f l f l b l / SIXTEEN           Align             1 / - -                l f l K l f l / corners
    '- - - - - - - - - -'  TWISTS     - - - - - - - - - - - - - -              '- - - - - - - - - -'  done
IVa(bv)   .- - - - - - - - - - -,      top done (#)      20 / 24               IVc(bv)   ,- - - - - - - - - - -,
        / e / K / r / l           ========================                    / b / r / b / l
      /- - -/- - -/- - -/ K l    (*) RR/0 ('r' face)     10 / - -              /- - -/- - -/- - -/ K :
      / r / b / f / l / l            R/B/1/1 ('r' face)   8 / - -               / b / b / f / l / l
     /- - -/- - -/- - -/ b / e :     R/B/1/1 ('r' face)   8 / - -              /- - -/- - -/- - -/ e / K :
    / r / f / t / l / l / l         - - - - - - - - - - - - - -                / b / r / b / l / l / l
    :- - - - - - - - - -: e / K / e :  corners done      26 / 26              :- - - - - - - - - -: K / K / K :
    l b l t l K l / l / l          ========================                   l e l b l e l / l / l
    :- - - - - - - - - -: r / t /                                             :- - - - - - - - - -: K / K /
    l r l e l K l / l /            Adjust (-1Qt)          1 / 1                l b l e l r l / l /
    :- - - - - - - - - -: r /       f,b (rdy)            - -/ - -              :- - - - - - - - - -: K /
    l b l e l t l / SIXTEEN         R/V/1/1 ('e' face)    7 / 3                l e l e l e l / corners
    '- - - - - - - - - -'  TWISTS   K,b (rdy)            - -/ - -              '- - - - - - - - - -'  done
IVb(tv)   .- - - - - - - - - - -,      R/V/1/1 ('r' face)  7 / 3               IVd(tv)   ,- - - - - - - - - - -,
        / t / t / t / l           Adjust (-1Qt)          1 / 1                    / t / t / t / l
      /- - -/- - -/- - -/ r :        e,b (rdy, Inv 'K')  1 / - -                /- - -/- - -/- - -/ r :
      / t / t / t / l / l            R/V/1/1 ('K' face)   7 / 3                  / t / t / t / l / l
     /- - -/- - -/- - -/ r / K :     Re-inv 'K'           1 / - -              /- - -/- - -/- - -/ r / f :
    / t / t / t / l / l / l          Adjust (-1Qt)        1 / 1                 / t / t / t / l / l / l
    :- - - - - - - - - -: r / r / K :  r,b (rdy, Inv 'f')  1 / - -             :- - - - - - - - - -: r / r / r :
    l f l f l f l / l / l           R/V/1/1 ('f' face)   7 / 3                 l f l f l f l / l / l
    :- - - - - - - - - -: b / e /     Re-inv 'f'          1 / - -             :- - - - - - - - - -: K / r /
    l e l f l f l / l /            Align (-1Qt)           1 / 1                l f l f l e l / l /
    :- - - - - - - - - -: e /       - - - - - - - - - - - - - -                :- - - - - - - - - -: r /
    l e l r l K l / top            bottom done           36 / 52              l f l f l f l / bottom
    '- - - - - - - - - -'  done    ========================                   '- - - - - - - - - -'  done
IVb(bv)   .- - - - - - - - - - -,      RM/0 ('f' face)    4 / 2                IVd(bv)   ,- - - - - - - - - - -,
        / b / f / r / l           Inv 'e'                1 / - -                   / b / b / b / l
      /- - -/- - -/- - -/ b :        ALG01 ('f' face)    45 / 22               /- - -/- - -/- - -/ K :
      / b / b / b / l / l            Re-inv 'e'           1 / - -                / b / b / b / l / l
     /- - -/- - -/- - -/ K / r :     - - - - - - - - - - - - - -               /- - -/- - -/- - -/ K / r :
    / f / e / b / l / l / l          edges done          51 / 75               / b / b / b / l / l / l
    :- - - - - - - - - -: f / K / K :  ====================                   :- - - - - - - - - -: K / K / K :
    l b l K l r l / l / l          CUBE DONE             133 / 177            l e l e l e l / l / l
    :- - - - - - - - - -: f / K /     ====================                    :- - - - - - - - - -: K / K /
    l b l e l r l / l /            (#) K,r,b and e,f,b in position             l e l e l r l / l /
    :- - - - - - - - - -: K /       (*) f,r,b is OK                            :- - - - - - - - - -: K /
    l e l e l e l / top                                                       l e l e l e l / bottom
    '- - - - - - - - - -'  done            Example 14                         '- - - - - - - - - -'  done
```

ADDENDUM

```
Va(tv)    .------------,            Cube/Op./Tr.      Stp/Tws      Vc(tv)    .------------,
       / r / k / e / |             t,e,f (OK)         --/--              / t / t / t / |
      /---/---/---/ k :            t,k                2 /--             /---/---/---/ r :
     / e / t / e /|/ |             t,e                4 / 2            / t / t / t /|/ |
    /---/---/---/ k / e :          t,f                4 /--           /---/---/---/ r / b :
   / t / e / r /|/|/ |             t,r                4 / 2          / t / t / t /|/|/ |
  :------------: f / r / e :       (t,f,r)            4 /--         :------------: r / r / r :
  | f | f | t |/|/|/ |             t,f,r              3 /--         | f | f | f |/|/|/ |
  :------------: k / r /           t,k,r              4 /--         :------------: b / r /
  | f | f | t |/|/ |              (t,k,e)             4 /--         | e | f | k |/|/ |
  :------------: t /               t,k,e              3 /--         :------------: r /
  | r | f | k |/SEVENTEEN Align    1 /--                            | f | f | f |/ corners
  '------------'  TWISTS           --------------------            '------------'  done
Va(bv)    .------------,           top done           33 / 31      Vc(bv)    .------------,
       / e / b / b / |             ===================                    / b / k / b / |
      /---/---/---/ f |            Align (+1Qt)  (#)   1 /--             /---/---/---/ k :
     / b / b / b /|/ |         (*)R/B/1/1 ('e' face)   8 /--           / r / b / e /|/ |
    /---/---/---/ k / t :          R/B/1/1 ('f face)   8 /--          /---/---/---/ f / r :
   / t / t / b /|/|/ |             R/B/1/1 ('f face)   8 /--         / b / b / b /|/|/ |
  :------------: r / k / b :       --------------------            :------------: k / k / k :
  | k | r | k |/|/|/ |             corners done       25 / 25       | e | f | e |/|/|/ |
  :------------: f / r /           ===================             :------------: e / k /
  | r | e | t |/|/ |               e,b (rdy)           --/--        | k | e | b |/|/ |
  :------------: f /               R/V/1/1 ('k' face)  7 / 3        :------------: k /
  | e | b | b |/SEVENTEEN          k,b (rdy)           --/--        | e | e | e |/ corners
  '------------'  TWISTS           R/V/1/1 ('r' face)  7 / 3        '------------'  done
Vb(tv)    .------------,           Adjust (+1Qt)       1 / 1        Vd(tv)    .------------,
       / t / t / t / |             f,b (rdy)           --/--              / t / t / t / |
      /---/---/---/ r :            R/V/1/1 ('e' face)  7 / 3             /---/---/---/ r :
     / t / t / t /|/ |             Adjust (+1Qt)       1 / 1           / t / t / t /|/ |
    /---/---/---/ r / b :          r,b (rdy, Inv 'f)   1 /--          /---/---/---/ r / e :
   / t / t / t /|/|/ |             R/V/1/1 ('f face)   7 / 3         / t / t / t /|/|/ |
  :------------: r / r / e :       Re-inv 'f           1 /--        :------------: r / r / r :
  | f | f | f |/|/|/ |                                              | f | f | f |/|/|/ |
  :------------: b / f /                                            :------------: k / r /
  | e | f | k |/|/ |               Align (2Qt's)       1 / 1        | f | f | r |/|/ |
  :------------: k /               --------------------            :------------: r /
  | r | r | r |/ top               bottom done        33 / 48      | f | f | f |/ bottom
  '------------' done              ===================             '------------' done
Vb(bv)    .------------,           RM/o ('f face)      4 / 2        Vd(bv)    .------------,
       / b / r / k / |             RM/o ('f face)      4 / 2              / b / b / b / |
      /---/---/---/ b :            ALG01 ('r' face)   45 / 22           /---/---/---/ k :
     / k / b / e /|/ |             --------------------              / b / b / b /|/ |
    /---/---/---/ f / r :          edges done         53 / 69        /---/---/---/ k / k :
   / b / b / f /|/|/ |             ===================             / b / b / b /|/|/ |
  :------------: b / k / k :       CUBE DONE         144 / 179     :------------: k / k / k :
  | f | f | e |/|/|/ |             ===================             | e | e | e |/|/|/ |
  :------------: e / k /           (#) all four bottom corners     :------------: r / k /
  | k | e | b |/|/ |                            in position        | e | e | f |/|/ |
  :------------: k /               (*) OOS1                        :------------: k /
  | e | e | e |/ top                     Example 15               | e | e | e |/ bottom
  '------------' done                                             '------------' done
```

ADDENDUM

```
VIa(tv)    .----------,          Cube/Op./Tr.      Stp/Tws      VIc(tv)    ,----------,
        / e / K / t / |          Align top (+1Qt)    1 /--              / t / t / t / |
     /---/---/---/ r :           t,f (OK)            --/--           /---/---/---/ r :
      / t / t / t / |/|          t,K (OK)            --/--            / t / t / t /|/|
   /---/---/---/ K / t :         t,e                  3 / 2        /---/---/---/ r / r :
    / f / b / e /|/|/|           t,f,r                3 /--         / t / t / t /|/|/|
 :----------: b / r / r :        t,f,e                3 /--      :----------: r / r / r :
 | b | r | K |/|/|/              t,r                  4 / 1      | f | f | f |/|/|/
 :----------: b / K /            (t,e,K)              3 /--      :----------: e / e /
 | f | f | f |/|/                t,e,K                3 /--      | r | f | K |/|/
 :----------: r /                t,K,r                4 /--      :----------: r /
 | t | t | f |/EIGHTEEN          Align (+1Qt)         1 /--      | f | f | f |/ corners
 '----------' TWISTS             --------------------           '----------' done

VIa(bv)  .----------,            top done (#)        25 / 28     VIc(bv)  ,----------,
        / b / b / b / |          ==================              / b / b / b / |
     /---/---/---/ K |           Align (+1Qt)  (@)    1 /--           /---/---/---/ K :
      / e / b / e /|/|       (*) R/O ('e' face)        9 /--          / e / b / b /|/|
   /---/---/---/ b / r :         R/B/1/1 ('e' face)    8 /--       /---/---/---/ K / K :
    / r / r / t /|/|/|           R/B/1/1 ('e' face)    8 /--        / b / b / b /|/|/|
 :----------: f / K / K :        --------------------           :----------: K / K / K :
 | f | f | e |/|/|/              corners done         26 / 26    | e | r | e |/|/|/
 :----------: r / e /            ==================              :----------: b / K /
 | e | e | K |/|/                K,b (OK)             --/--       | f | e | f |/|/
 :----------: t /                f,b (rdy, Inv 'e')    1 /--      :----------: K /
 | e | f | K |/EIGHTEEN          R/V/1/1 ('e' face)    7 / 3      | e | e | e |/ corners
 '----------' TWISTS             Re-inv 'e'            1 /--      '----------' done

VIb(tv)  .----------,            [R/V/1/1 ('f' face)]  7 / 3      VId(tv)  ,----------,
        / t / t / t / |          e,b (rdy)            --/--              / t / t / t / |
     /---/---/---/ r :           R/V/1/1 ('K' face)    7 / 3           /---/---/---/ r :
      / t / t / t /|/|           r,b (rdy)            -- /--            / t / t / t /|/|
   /---/---/---/ r / r :         R/V/1/1 ('f' face)    7 / 3        /---/---/---/ r / K :
    / t / t / t /|/|/|                                              / t / t / t /|/|/|
 :----------: r / r / K :                                        :----------: r / r / r :
 | f | f | f |/|/|/                                              | f | f | f |/|/|/
 :----------: e / K /                                            :----------: e / r /
 | r | f | K |/|/                                                | K | f | f |/|/
 :----------: r /                                                :----------: r /
 | b | b | f |/ top              bottom done          30 / 42    | f | f | f |/ bottom
 '----------' done               ==================              '----------' done

VIb(bv)  .----------,                                            VId(bv)  ,----------,
        / b / b / b / |                                                  / b / b / b / |
     /---/---/---/ e :           RM/O ('K' face)       4 / 2           /---/---/---/ K :
      / r / b / b /|/|           ALG01 ('r' face)     45 / 22           / b / b / b /|/|
   /---/---/---/ e / K :         --------------------             /---/---/---/ K / r :
    / r / f / f /|/|/|           edges done           49 / 73      / b / b / b /|/|/|
 :----------: b / K / K :        ==================              :----------: K / K / K :
 | K | e | e |/|/|/              CUBE DONE          130 / 169    | e | e | e |/|/|/
 :----------: b / K /            ==================              :----------: r / K /
 | f | e | f |/|/                (#) f,r,b is OK                  | e | e | f |/|/
 :----------: K /                (@) e,K,b and e,f,b in position  :----------: K /
 | e | e | e |/ top              (*) e,K,b is OK                  | e | e | e |/ bottom
 '----------' done                    Example 16                 '----------' done
```

ADDENDUM

```
VIIa(tv)  .----------,        Cube/Op./Tr.    Stp/Tws    VIIc(tv)  .----------,
          / t / k / f / |        t,e,k (OK)      --/--              / t / t / t / |
         /---/---/---/ t :         t,e            3 / 1            /---/---/---/ r :
        / e / t / f / | /|        Align top       1 /--           / t / t / t / | /|
       /---/---/---/ e / t :       t,k,r          3 /--          /---/---/---/ r / r :
      / b / f / f / | / | /|        t,k           3 / 2          / t / t / t / | / | /|
     :----------: b / r / f :       t,r           3 / 1         :----------: r / r / r :
     | k | t | e | / | / | /        t,f,e         2 /--         | f | f | f | / | / | /
     :----------: b / k /          (t,f,r)        4 /--         :----------: f / e /
     | t | f | f | / | /            t,f,r         3 /--         | b | f | b | / | /
     :----------: k /               t,f           3 / 2        :----------: r /
     | f | r | t | /NINETEEN        Align          1 / 1       | f | f | f | / corners
     '----------' TWISTS                                       '----------' done
                                    ----------------------
VIIa(bv) .----------,               top done (#)     26 / 33   VIIc(bv) .----------,
         / r / e / r / |            ==================          / b / f / b / |
        /---/---/---/ t |                                      /---/---/---/ k :
       / b / b / k / | /|       (*) R/B/1/1 ('k' face)  8 /--   / r / b / b / | /|
      /---/---/---/ t / e :         R/B/1/1 ('r' face)  8 /--  /---/---/---/ e / k :
     / r / b / r / | / | /|         R/B/1/1 ('r' face)  8 /--  / b / b / b / | / | /|
    :----------: k / k / e :        ----------------------    :----------: k / k / k :
    | b | k | b | / | / | /         corners done    24 / 24    | e | r | e | / | / | /
    :----------: f / r /            ==================         :----------: e / k /
    | r | e | r | / | /             Adjust (2Qt's)    1 / 1     | k | e | k | / | /
    :----------: k /                b,k (rdy)        --/--     :----------: k /
    | e | b | e | /NINETEEN         R/V/1/1 ('r' face)  7 / 3   | e | e | e | / corners
    '----------' TWISTS             Adjust (-1Qt)     1 / 1    '----------' done
VIIb(tv) .----------,               e,b (rdy)        --/--      VIId(tv) .----------,
         / t / t / t / |            R/V/1/1 ('k' face)  7 / 3           / t / t / t / |
        /---/---/---/ r :           r,b (rdy)        --/--            /---/---/---/ r :
       / t / t / t / | /|           R/V/1/1 ('f' face)  7 / 3        / t / t / t / | /|
      /---/---/---/ r / r :         Adjust (2Qt's)    1 / 1         /---/---/---/ r / k :
     / t / t / t / | / | /|         f,b (rdy, Inv 'e')  1 /--       / t / t / t / | / | /|
    :----------: r / r / k :        R/V/1/1 ('e' face)  7 / 3      :----------: r / r / r :
    | f | f | f | / | / | /         Re-inv 'e'        1 /--        | f | f | f | / | / | /
    :----------: f / e /            Align (+1Qt)      1 / 1        :----------: r / r /
    | b | f | b | / | /                                           | e | f | k | / | /
    :----------: f /                ----------------------        :----------: r /
    | f | e | b | / top             bottom done     34 / 50       | f | f | f | / bottom
    '----------' done               ==================            '----------' done
VIIb(bv) .----------,                                             VIId(bv) .----------,
         / r / b / r / |            RM/0 ('f' face)   4 / 2                / b / b / b / |
        /---/---/---/ b :           RM/0 ('f' face)   4 / 2               /---/---/---/ k :
       / f / b / r / | /|           ALG01 ('e' face) 45 / 22            / b / b / b / | /|
      /---/---/---/ f / k :         ----------------------             /---/---/---/ k / e :
     / b / b / b / | / | /|         edges done       53 / 79           / b / b / b / | / | /|
    :----------: k / k / k :        ==================                :----------: k / k / k :
    | e | r | e | / | / | /         CUBE DONE       137 / 186          | e | e | e | / | / | /
    :----------: e / k /            ==================                :----------: f / k /
    | k | e | k | / | /             (#) e,k,b and e,f,b are OK         | f | e | r | / | /
    :----------: k /                (*) OOS2                           :----------: k /
    | e | e | e | / top                                               | e | e | e | / bottom
    '----------' done                  Example 17                     '----------' done
```

```
VIIIa(tv)  .-----------,
         / b / e / b / |
       /---/---/---/ k :
      / r / t / t /|/ |
    /---/---/---/ f / e :
   / t / k / t /|/|/ |
 :-----------: f / r / e :
 | e | b | e |/|/|/ |
 :-----------: b / k /
 | f | f | r |/|/
 :-----------: k /
 | r | f | b |/ TWENTY-
 '-----------'ONE TWISTS

VIIIa(bv)  .-----------,
         / e / e / b / |
       /---/---/---/ f |
      / e / b / t /|/ |
    /---/---/---/ k / b :
   / t / r / r /|/|/ |
 :-----------: f / k / r :
 | k | t | t |/|/|/ |
 :-----------: f / t /
 | r | e | b |/|/
 :-----------: r /
 | k | k | f |/ TWENTY-
 '-----------'ONE TWISTS

VIIIb(tv)  .-----------,
         / t / t / t / |
       /---/---/---/ r :
      / t / t / t /|/ |
    /---/---/---/ r / f :
   / t / t / t /|/|/ |
 :-----------: r / r / k :
 | f | f | f |/|/|/ |
 :-----------: k / k /
 | e | f | r |/|/
 :-----------: e /
 | r | k | b |/ top
 '-----------' done

VIIIb(bv)  .-----------,
         / f / b / r / |
       /---/---/---/ b :
      / e / b / r /|/ |
    /---/---/---/ b / b :
   / b / r / b /|/|/ |
 :-----------: k / k / k :
 | f | f | e |/|/|/ |
 :-----------: e / k /
 | f | e | b |/|/
 :-----------: k /
 | e | e | e |/ top
 '-----------' done
```

Cube/Op./Tr.	Stp/Tws
Align top	1/--
t,f,e (OK)	--/--
t,f (OK)	--/--
t,e,k (OK)	--/--
t,k	3/--
t,f,,r	4/--
t,r	4/1
t,e	4/2
(t,k,r)	3/--
t,k,r	4/--
Align	1/--
-------------------	---
top done (#)	24 / 27
====================	
R/O ('k' face)	9/--
(*)R/B/1/1 ('k' face)	8/--
R/B/1/1 ('e' face)	8/--
R/B/1/1 ('e' face)	8/--
-------------------	---
corners done	33 / 33
====================	
e,b (rdy)	--/--
R/V/1/1 ('k' face)	7/3
f,b (rdy)	--/--
R/V/1/1 ('e' face)	7/3
Adjust (2Qt's)	1/1
k,b (rdy)	--/--
R/V/1/1 ('r' face)	7/3
r,b (rdy, Inv 'f')	1/--
R/V/1/1 ('f' face)	7/3
Re-inv 'f'	1/--
Align (2Qt's)	1/1
-------------------	---
bottom done	32 / 46
====================	
RM/O ('r' face)	4/2
Inv 'e'	1/--
ALG01 ('f' face)	45 / 22
Re-inv 'e'	1/--
-------------------	---
edges done	51 / 75
====================	
CUBE DONE	140 / 181
====================	
(#) k,r,b and k,e,b in position	
(*) OOSI	

Example 18

```
VIIIc(tv)  .-----------,
         / t / t / t / |
       /---/---/---/ r :
      / t / t / t /|/ |
    /---/---/---/ r / f :
   / t / t / t /|/|/ |
 :-----------: r / r / r :
 | f | f | f |/|/|/ |
 :-----------: k / r /
 | e | f | r |/|/
 :-----------: r /
 | f | e | f |/ corners
 '-----------' done

VIIIc(bv)  .-----------,
         / b / f / b / |
       /---/---/---/ k :
      / k / b / r /|/ |
    /---/---/---/ b / b :
   / b / b / b /|/|/ |
 :-----------: k / k / k :
 | e | k | e |/|/|/ |
 :-----------: e / k /
 | f | e | b |/|/
 :-----------: k /
 | e | e | e |/ corners
 '-----------' done

VIIId(tv)  .-----------,
         / t / t / t / |
       /---/---/---/ r :
      / t / t / t /|/ |
    /---/---/---/ r / e :
   / t / t / t /|/|/ |
 :-----------: r / r / r :
 | f | f | f |/|/|/ |
 :-----------: f / r /
 | e | f | r |/|/
 :-----------: r /
 | f | f | f |/ bottom
 '-----------' done

VIIId(bv)  .-----------,
         / b / b / b / |
       /---/---/---/ k :
      / b / b / b /|/ |
    /---/---/---/ k / f :
   / b / b / b /|/|/ |
 :-----------: k / k / k :
 | e | e | e |/|/|/ |
 :-----------: k / k /
 | k | e | r |/|/
 :-----------: k /
 | e | e | e |/ bottom
 '-----------' done
```

ADDENDUM

```
iXa(tv)  .-----------,              Cube/Op./Tr.    Stp/Tws      IXc(tv)  ,-----------,
       / t / f / t / |              Align top (-1Qt)   1 /--            / t / t / t / |
      /---/---/---/ k :               t,e,k (OK)       --/--          /---/---/---/ r :
     / k / t / r /|/|                  t,r             2 /--          / t / t / t /|/|
    /---/---/---/ f / r :              t,r,k (OK)      --/--         /---/---/---/ r / r :
   / k / f / e /|/|/|                  t,f             2 /--        / t / t / t /|/|/|
  :-----------: t / r / r :            t,f,,e (OK)     --/--       :-----------: r / r / r :
  | r | b | f |/|/|/|                   t,e            3 / 2       | f | f | f |/|/|/|
  :-----------: t / e /                (t,f,r)         3 /--       :-----------: b / k /
  | r | f | f |/|/|                     t,f,r          3 /--       | b | f | f |/|/|
  :-----------: b /                     t,k            4 / 1       :-----------: r /
  | k | k | r |/ TWENTY-                                          | f | e | f |/ corners
  '-----------'THREE TWISTS                                       '-----------' done

iXa(bv)  .-----------,              top done (#)     18 / 21      IXc(bv)  ,-----------,
       / f / b / b / |              ====================               / b / e / b /|
      /---/---/---/ k |             (@)Align (+1Qt)    1 /--          /---/---/---/ k :
     / b / b / k /|/|                 R/0 ('e' face)   9 /--          / f / b / r /|/|
    /---/---/---/ r / b :          (*)R/B/1/1 ('e' face) 8 /--       /---/---/---/ k / b :
   / e / t / b /|/|/|                 R/B/1/1 ('f' face) 8 /--      / b / f / b /|/|/|
  :-----------: e / k / e :          --------------------          :-----------: k / k / k :
  | b | k | f |/|/|/|                corners done      26 / 26      | e | r | e |/|/|/|
  :-----------: t / e /             ====================           :-----------: k / k /
  | t | e | e |/|/                  Adjust (-1Qt)      1 / 1        | e | e | b |/|/|
  :-----------: f/                  f,b (rdy)          --/--        :-----------: k /
  | t | e | r |/ TWENTY-            R/V/1/1 ('e' face)  7 / 3       | e | e | e |/ corners
  '-----------'THREE TWISTS                                        '-----------' done

iXb(tv)  .-----------,              Adjust (+1Qt)      1 / 1        IXd(tv)  ,-----------,
       / t / t / t / |              r,b (rdy)          --/--              / t / t / t / |
      /---/---/---/ r :             R/V/1/1 ('f' face)  7 / 3          /---/---/---/ r :
     / t / t / t /|/|               k,b (rdy)          --/--           / t / t / t /|/|
    /---/---/---/ r / r :           R/V/1/1 ('r' face)  7 / 3         /---/---/---/ r / e :
   / t / t / t /|/|/|               e,b (rdy)          --/--         / t / t / t /|/|/|
  :-----------: r / r / b :         R/V1/1 ('k' face)   7 / 3        :-----------: r / r / r :
  | f | f | f |/|/|/|                                                | f | f | f |/|/|/|
  :-----------: b / k /                                              :-----------: f / r /
  | b | f | f |/|/|                                                  | e | f | r |/|/|
  :-----------: r /                 --------------------             :-----------: r /
  | b | r | f |/ top                bottom done       30 / 44        | f | f | f |/ bottom
  '-----------' done                ====================            '-----------' done

iXb(bv)  .-----------,              RM/0 ('r' face)    4 / 2        IXd(bv)  ,-----------,
       / b / e / e / |              Inv 'e'            1 /--              / b / b / b / |
      /---/---/---/ k :             ALG01 ('f' face)  45 / 22         /---/---/---/ k :
     / k / b / f /|/|               Re-inv 'e'         1 /--          / b / b / b /|/|
    /---/---/---/ r / b :           --------------------            /---/---/---/ k / f :
   / r / e / b /|/|/|               edges done        51 / 75       / b / b / b /|/|/|
  :-----------: e / k / k :         ====================            :-----------: k / k / k :
  | k | f | f |/|/|/|               CUBE DONE         125 / 166     | e | e | e |/|/|/|
  :-----------: k / k /             ====================            :-----------: k / k /
  | e | e | b |/|/                  (#) f,r,b is OK                  | k | e | r |/|/|
  :-----------: k /                 (*) see page (17) "in-sync"      :-----------: k /
  | e | e | e |/ top                (@) r.b.f & r,b,k in position    | e | e | e |/ bottom
  '-----------' done                      Example 19                '-----------' done
```

```
Xa(tv)     .-----------,
         / e  / k / t /|
        /---/---/---/ r :
       / b / t / b /|/|
      /---/---/---/ r / b :
     / t / f / e /|/|/|
     :-----------: b/r/r:
     | e | t | k |/|/|/
     :-----------: f/k/
     | t | f | e |/|/
     :-----------: r /
     | f | f | f |/ TWENTY-
     '-----------'FOUR TWISTS

Xa(bv)     .-----------,
         / b / e / b /|
        /---/---/---/ k |
       / b / b / k /|/|
      /---/---/---/ r / e :
     / t / t / f /|/|/|
     :-----------: b/k/k:
     | r | e | e |/|/|/
     :-----------: r / t /
     | r | e | f |/|/
     :-----------: f /
     | k | k | t |/ TWENTY-
     '-----------'FOUR TWISTS

Xb(tv)     .-----------,
         / t / t / t /|
        /---/---/---/ r :
       / t / t / t /|/|
      /---/---/---/ r / r :
     / t / t / t /|/|/|
     :-----------: r/r/e:
     | f | f | f |/|/|/
     :-----------: e / f /
     | r | f | f |/|/
     :-----------: r /
     | f | r | b |/ top
     '-----------' done

Xb(bv)     .-----------,
         / k / b / k /|
        /---/---/---/ b :
       / f / b / b /|/|
      /---/---/---/ k / b :
     / r / e / e /|/|/|
     :-----------: f/k/k:
     | b | k | b |/|/|/
     :-----------: b / k /
     | k | e | e |/|/
     :-----------: k /
     | e | e | e |/ top
     '-----------' done
```

Cube/Op./Tr.	Stp/Tws
t,r,k (OK)	--/--
t,e	3 / 1
t,e,k	3 /--
t,r	3 / 1
t,e,f	2 /--
t,r,f	3 /--
t,f	4 / 2
(t,k)	3 / 2
t,k	4 / 2
Align	1 /--
-----------------------	
top done	26 / 33
=====================	
Align (+1Qt) (#)	1 /--
(*) R/B/1/1 ('r' face)	8 /--
R/B/1/1 ('k' face)	8 /--
----------------------	
corners done	17 / 17
=====================	
Adjust (-1Qt)	1 / 1
r,b (rdy)	--/--
R/V/1/1 ('f' face)	7 / 3
Adjust (+1Qt)	1 / 1
f,b (rdy)	--/--
R/V/1/1 ('e' face)	7 / 3
k,b (rdy)	--/--
R/V/1/1 ('r' face)	7 / 3
e,b (rdy, Inv 'k')	1 /--
R/V1/1 ('k' face)	7 / 3
Re-inv 'k'	1 /--
----------------------	
bottom done	32 / 46
=====================	
RM/0 ('f' face)	4 / 2
ALG01 ('r' face)	45 / 22
----------------------	
edges done	49 / 73
=====================	
CUBE DONE	124 / 169
=====================	

(#) all four bottom corners
               in position
(*) see page 17 "in-sync"
          Example 20

```
Xc(tv)      .-----------,
          / t / t / t /|
         /---/---/---/ r :
        / t / t / t /|/|
       /---/---/---/ r / r :
      / t / t / t /|/|/|
      :-----------: r/r/r:
      | f | f | f |/|/|/
      :-----------: e / f /
      | r | f | f |/|/
      :-----------: r /
      | f | k | f |/ corners
      '-----------' done

Xc(bv)      .-----------,
          / b / b / b /|
         /---/---/---/ k :
        / b / b / f /|/|
       /---/---/---/ r / b :
      / b / e / b /|/|/|
      :-----------: k/k/k:
      | e | k | e |/|/|/
      :-----------: b / k /
      | k | e | e |/|/
      :-----------: k /
      | e | e | e |/ corners
      '-----------' done

Xd(tv)      .-----------,
          / t / t / t /|
         /---/---/---/ r :
        / t / t / t /|/|
       /---/---/---/ r / f :
      / t / t / t /|/|/|
      :-----------: r/r/r:
      | f | f | f |/|/|/
      :-----------: e / r /
      | f | f | k |/|/
      :-----------: r /
      | f | f | f |/ bottom
      '-----------' done

Xd(bv)      .-----------,
          / b / b / b /|
         /---/---/---/ k :
        / b / b / b /|/|
       /---/---/---/ k / r :
      / b / b / b /|/|/|
      :-----------: k/k/k:
      | e | e | e |/|/|/
      :-----------: r / k /
      | e | e | k |/|/
      :-----------: k /
      | e | e | e |/ bottom
      '-----------' done
```

BRIEF SUMMARY
(EX11...EX20)

Exercise Eleven: After a R/O operation repositioned two adjacent bottom corners, none of them were okay, so a particular R/B/I/I translation got them 'in-sync' (see text, page 17) otherwise a random translation has only a 50-50 chance and then the OOS1, OOS2, or OOS3 routines are needed. In EX4, EX6, EX7, and EX10 a random R/B/I/I translation was done. An extra R/V/I/I translation was needed to bump back one bottom edge cube so it could be placed in a READY position.

Exercise Twelve: After a R/O operation repositioned two adjacent bottom corners, the four bottom corners were already 'in-sync'. All four bottom edge cubes were in the bottom layer incorrectly then the first R/V/I/I translation had to be a bumping translation. This is the first exercise that this has happen. After the four middle edge cubes were positioned they were also 'okay', no ALGO1 operation was necessary. This also happen in EX3 and EX5.

Exercise Thirteen: After a R/O operation repositioned two adjacent bottom corners, the four bottom corners were already 'in-sync'. One bottom edge cube was already okay but another edge cube had to be bumped back in order for it to be put in a READY position.

Exercise Fourteen: Two bottom corners were positioned diagonally, they needed a RR/O operation. This also happen in EX1. After the RR/O operation, the bottom four corners were already 'in-sync'. Two middle edge cubes that were okay were on a diagonal so a side needed to be inverted prior to an ALGO1 operation. This also happened in EX1 and EX7.

Exercise Fifteen: After an alignment all four of the bottom corners were in position so there was no need for either a R/O or a RR/O operation. This also happen in EX8 and EX9. An OOS1 (out-of-sync-1) routine got the four bottom four corners 'in-sync'. This OOS1 routine was also used in EX4.

Exercise Sixteen: After a R/O operation repositioned two adjacent bottom corners, the four bottom corners were already 'in-sync'. One bottom edge cube was already okay but another edge cube had to be bumped back in order for it to be put in a READY position.

Exercise Seventeen: All four of the bottom corners were in position so neither a R/O nor a RR/O operation was needed. This also happen in EX8, ES9, and EX15. An OOS2 (out-of-sync-2) routine got the four bottom corners 'in-sync'. This OOS2 routine was also used in EX6.

Exercise Eighteen: An OOS1 (out-of sync-1) routine got the bottom four corners 'in-sync'. This OOS1 routine was also used in EX4 and EX15. Two middle edge cubes that were okay were on a diagonal so a side needed to be inverted prior to an ALGO1 operation. This also happened in EX1, EX7 and EX14.

Exercise Nineteen: After a R/O operation repositioned two adjacent bottom corners, none of them were okay, so a particular R/B/I/I translation got them 'in-sync' as for EX11 (see EX11 brief summary). Two middle edge cubes that were okay were on a diagonal so a side needed to be inverted prior to an ALGO1 operation. This also happened in EX1, EX7, EX14 and EX18.

Exercise Twenty: After an alignment all four bottom corners were in position but none of them were okay, so a particular R/B/I/I translation got them 'in-sync' as for EX11 and EX19 (see EX11 brief summary).

Exercise PPS: Like EX1 and EX9 an OOS3 (out-of-sync-3) routine got the four bottom corners 'in-sync'. Like EX12 the first R/V/I/I translation was a bumping operation. Like EX2 and EX8 a C/V/I/I translation positioned the four middle edge cubes.

ERRATUM
While authenticating the eighty figures of EX11 through EX20 using the CUBIK solution in reverse, two figures had errors in the manuscript as follows:
1) the right side of Figure IIIb(bv) was shifted ninty degrees, is now r,b,k; k,k,e; k,k,k; was k,k,r; k,k,b; k,e,k, and,
2) Figure IVc(tv) had a r,r edge cube, however, by deduction should be edge cube r,f.

ADDENDUM

TABLE 10EX'

| tws----12 | 13 | 14 | 16 | 17 | 18 | 19 | 21 | 23 | 24 | eyes- |
path EX11	EX12	EX13	EX14	EX15	EX16	EX17	EX18	EX19	EX20	closed
i	X	X		X	X				X	X	0 or 1 ?
ii	X			X			X	X			2 or more?
iii					X		X			X	4 pos. ?
iv	X	X	X			X		X	X	X	R/o
v				X							RR/o
stp	31	19	29	20	33	25	26	24	18	26	cor. pos.
tws	37	27	36	24	37	28	33	27	21	33	done
vi	X								X	X	one OK? #
vii	X	X	X	X	X	X	X	X	X	X	in-sync1 #
vii		X		X	X	X	X	X			in-sync2 #
viii'					X			X			OOS1 #
viii''							X				OOS2 #
viii'''											OOS3 #
=========	==	==	==	==	==	==	==	==	==	=corns. OK
ix	X	X	X	X	X		X	X	X	X	1st bot.
lx	X	X	X	X	X	X	X	X	X	X	2nd edge
lx	X	X	X	X	X	X	X	X	X	X	3rd OK
lx	X	X	X	X	X	X	X	X	X	X	4th *
x		X									1st B
x											2nd U
x						X					3rd M
x	X		X								4th P*
xi											1st inv.
xi	X				X						2nd inv.
xi	X			X							3rd inv.
xi	X	X	X	X	X		X	X		X	4th inv.
=========	==	==	==	==	==	==	==	==	==	= bot. OK
xii	X	X	X	X	X	X	X	X	X	X	one pos.?
xiii											on diag. @
xiv											on side @
xv								X			two pos.?
xvi	X	X	X	X	X	X	X		X	X	RM/0-1
xvi		X	X		X		X				RM/0-2
=========	==	==	==	==	==	==	==	==	==	=mid. Pos. done
xvii		X									four OK?
xviii				X				X	X		inv. <>
xix	X		X		X	X	X			X	once <>
xix											twice <>
=========	==	==	==	==	==	==	==	==	==	=DONE
steps	150	95	133	133	144	130	137	140	125	124	(1311)
twists	199	126	181	177	179	169	186	181	166	169	(1733)

[average steps = 131 per exercise; average twists = 173 per exercise]
[minimum steps is 95; maximum steps is 150]

= R/B/1/1   * = R/V/1/1   @ = C/V/1/2   <> = ALG01

[average steps top face done = 25 per exercise; average twists top face done = 30 per exercise]
[minimum steps top face is 18; maximum steps top face is 33]

ADDENDUM

```
PPSa(tv) .------------,        Cube/Op./Tr.   Stp/Tws     PPSc(tv) .------------,
         / t / r / f /|                                            / t / t / t /|
        /---/---/---/ e :       t,e,k (OK)     --/--              /---/---/---/ r :
       / e / t / r /|/|          t,f,r          2 /--            / t / t / t /|/|
      /---/---/---/ k / t :      t,r (OK)       --/--           /---/---/---/ r / k :
     / b / f / f /|/|/|          t,e            3 / 2          / t / t / t /|/|/|
     :-----------: e / r / r :   t,k            3 / 2          :-----------: r / r / r :
     | f | t | t |/|/|/|         t,f,e          3 /--          | f | f | f |/|/|/|
     :-----------: t / k /       (t,k,r)        4 /--          :-----------: e / b /
     | e | f | k |/|/            t,k,r          3 /--          | r | f | f |/|/
     :-----------: r /          (t,f)           3 / 2          :-----------: r /
     | f | b | b |/ NINE         t,f            4 / 2          | f | b | f |/ corners
     '-----------' TWISTS      ----------------------          '-----------' done
PPSa(bv) .------------,         top done (#)   25 / 33     PPSc(bv) ,------------,
         / k / e / k /|        ======================               / b / e / b /|
        /---/---/---/ t |       Align 'b' (2Qt's) (#)  1 /--        /---/---/---/ k :
       / k / b / b /|/|         R/O ('e' face)   9 /--            / k / b / r /|/|
      /---/---/---/ f / e :  (*) R/B/1/1 ('f face)  8 /--        /---/---/---/ b / e :
     / t / t / b /|/|/|         R/B/1/1 ('k' face)  8 /--       / b / b / b /|/|/|
     :-----------: k / k / b :  R/B/1/1 ('k' face)  8 /--       :-----------: k / k / k :
     | r | r | e |/|/|/|       ----------------------           | e | f | e |/|/|/|
     :-----------: r / b /      corners done   34 / 34          :-----------: r / k /
     | f | e | f |/|/          ======================           | k | e | f |/|/
     :-----------: k /        (@) [R/V/1/1 ('f' face)]  7 / 3    :-----------: k /
     | r | b | e |/ NINE       Adjust (-1Qt)  1 / 1             | e | e | e |/ corners
     '-----------' TWISTS      e,b (rdy, Inv 'k')  1 /--        '-----------' done
PPSb(tv) .------------,        R/V/1/1 ('k' face)  7 / 3     PPSd(tv) ,------------,
         / t / t / t /|        Re-inv 'k'  1 /--                      / t / t / t /|
        /---/---/---/ r :      Adjust (-1Qt)  1 / 1                  /---/---/---/ r :
       / t / t / t /|/|        f,b (rdy)  --/--                    / t / t / t /|/|
      /---/---/---/ r / k :    R/V/1/1 ('e' face)  7 / 3          /---/---/---/ r / f :
     / t / t / t /|/|/|        Adjust (-1Qt)  1 / 1              / t / t / t /|/|/|
     :-----------: r / r / b :  k,b (rdy, Inv 'r')  1 /--        :-----------: r / r / r :
     | f | f | f |/|/|/|       R/V/1/1 ('r' face)  7 / 3         | f | f | f |/|/|/|
     :-----------: e / f /     Re-inv 'r'  1 /--                 :-----------: k / r /
     | r | f | f |/|/          r,b (rdy, Inv 'f')  1 /--         | k | f | e |/|/
     :-----------: e /         R/V/1/1 ('f' face)  7 / 3         :-----------: r /
     | f | b | k |/ top        Re-inv 'f'  1 /--                 | f | f | f |/ bottom
     '-----------' done        Align (-1Qt)  1 / 1               '-----------' done
PPSb(bv) .------------,        ----------------------        PPSd(bv ,------------,
         / b / b / f /|        bottom done  45 / 64                  / b / b / b /|
        /---/---/---/ e :      ======================                /---/---/---/ k :
       / e / b / b /|/|        C/V/1/2 (any face)  8 / 8           / b / b / b /|/|
      /---/---/---/ k / e :    ALG01 ('e' face)  45 / 22          /---/---/---/ k / e :
     / r / b / k /|/|/|        ----------------------            / b / b / b /|/|/|
     :-----------: b / k / k :  edges done  53 / 83              :-----------: k / k / k :
     | b | r | r |/|/|/|       ======================            | e | e | e |/|/|/|
     :-----------: r / k /      CUBE DONE  157 / 214             :-----------: f / k /
     | k | e | f |/|/          ======================            | r | e | r |/|/
     :-----------: k /        (#) e,k,b and e,f,b in position; (*) OOS3  :-----------: k /
     | e | e | e |/ top       (@) edges on bottom incorrectly    | e | e | e |/ bottom
     '-----------' done        Example 'PPS' ('extra')           '-----------' done
```

ADDENDUM

S U M M A R Y

The first figure (or pattern) appearing in the twenty examples is a scrambled CUBE as result of the following number of twists for each of the examples starting from a solid (pristine) CUBE: 10 through 25, 30, 33, 35, and 40. The average number of steps and twists are 137 steps and 181 twists using the CUBIK solution for the twenty examples tabulated in the two tables. However, the average number of steps for Exercises 1 through 10 are eleven more than the average for Exercises 11 through 20. As shown by a comparison table this was due to the fact that I was more efficient in solving the top layer for the last ten exercises and the first ten has one more ALGO1 operation which both account for an eighty step difference that is the major portion of the 110 total steps difference.

COMPARISON TABLE

Translations/Operations	EX1-10	EX11-20	delta EX1-10(steps)
R/0	7	6	+9
RR/0	1	1	---
R/B/1/1	24	22	+16
R/V/1/1 (move)	37	38	-7
R/V/1/1 (bump)	4	4	---
C/V/1/2	2	---	+16
RM/0	13	14	-4
ALGO1	10	9	+45
Top layer	286 steps	251 steps	+35
===			
TOTAL	1421 steps	1311 steps	+110

As you might imagine how I have enjoyed "cutting and pasting" (moving) a sundry of CUBES using the word processor in 1982 and recently in 2002. I was getting good at moving files all about too. This appeared so enjoyable to my son in law as he on occasion peered over my shoulder at a distance mumbled that I would never finish. It's very highly probable why I did the last ten exercises improving the nomenclature in order to do a KIBUC solution (for authentication purposes) on these last ten exercises, going from a solid CUBE back to a patterns Ib through Xb ("top done"). However, I made no attempt to go from patterns Ib through Xb ("top done") to patterns Ia through Xa (scrambled CUBE) respectively or vice versa. Patterns Ia through Xa were re-authenticated by reproducing them using the CUBIK solution. At last, REJOICE, I'm finish!

NOTE: As can be expected, statistically, when unscrambling the CUBE, infrequently the CUBE appears solved prematurely but with two cubes not oriented or three cubes out of position and not oriented. Then making the two cubes adjacent an ALGO1 operation finishes the CUBE earlier or putting the three cubes on one slice (when possible) one or two C/B/1/2 translations finishes the CUBE earlier.

ADDENDUM

PS: Now I will think about my next book "Winning Streaks" after taking data on 196,000 hands of B.J. (to date I have completed 132,000).

PS-PS My eleven-year old grandson wants to say "I hope you like my book and I hope you buy my sequel, BYE".

PS-PS-PS: At the very end of my manuscript I again bogged down trying to add headers and footers so again thanks to my son in law for helping me out. When it's all said and done I supposed I was so fixed on doing the "micro-processing" the 1982 way that I approached Microsoft Word 6.0 with a closed mind. I wasn't too friendly most of the time and should apologize for Microsoft Word 6.0 is forgiving after all.

PS-PS-PS-PS: My four year old grand daughter can't wait to color it!

> *Finally as this manuscript is soon becoming a galley it was unexpected the way the Print Data File embellish on the ANDY font, it was neat. But I still have to live with the cubic figure sketchiness, however, my son-in-law stipulated that that is my "artistic license"; I have to let him explain that one!*

ADDENDUM

(Positions Bottom Four Corners)

BEFORE				AFTER		
fix		y		fix		x
x		FIX		y		FIX

front   top views   front

R/0//B/0//Bk/0 = RR/0

BEFORE				AFTER		
v		w		w		v
FIX		fix		FIX		fix

front   top views   front

R/0//Bk/0 = R/0

* * * * * *

(Orients Bottom Four Corners)

tws		tws	
FIX		tws	

front

R/B/1/1

(ENDING Inverts j and k)

j		k	

front face

ALGO #1

* * * * * *

(Positions/Orients Bottom Four Edge Cubes)

BEFORE   AFTER

f | | | | f | | | |
r | | | | r | | | |
o | u | | | o | | | |
n | | | | n | | | |
t | | | | t | | u | |

views of right face

R/V/1/1

FIX = stays positioned and oriented
Q = ¼ turn   tws = twist
(?) = cube inverted   (ALGO #1)

1) R/V/1/1, (k) to bottom   5) R/V/1/1, send down 1st
2) R/V/1/1, (j) to bottom   6) Middle plus 2Qt's
3) C/B/1/2, inverts j & k   7) R/V/1/1, send down 2nd.
4) Middle minus 1Qt   8) R/V/1/1, send down 3rd

* * * * * * * * * * * *

(Positions Middle Edge Cubes)

BEFORE				AFTER		
m		n		p		o
o		p		n		m

front   top views   front

C/V/1/2

BEFORE				AFTER		
t		s		r		t
FIX		r		FIX		s

front   top views   front

R/0//R/0//M/0 = RM/0

* * * * * *

(Shifts Three Bottom Edge Cubes)

BEFORE   AFTER

	i				(h)	
h		g		g		(i)
	FIX				FIX	

front   top views   front

C/B/1/2

* * * * * *

*(Gets Bottom Four Corners In-Sync @R/B/1/1)

OOS1   b   OOS2

ok		ok				ok
						ok

b front b   top views   b front

(When two corners are OK)

ok			<b
			O
			O
		ok	S

b front   3

CUBIK Algorithms

www.ingramcontent.com/pod-product-compliance
Lightning Source LLC
Chambersburg PA
CBHW081123170526
45165CB00008B/2533

9 781418 454807